COMPLETE

AZ

Geography

Coursework

HANDBOOK

Geog

Coursework

HANDBOOK

Malcolm Skinner
David Redfern
Geoff Farmer

3rd edition

Hodder & Stoughton

A MEMBER OF THE HODDER HEADLINE GROUP

Orders: please contact ˸o˸˸po˸˸ Ltd, 130 Milton Park, Abingdon, Oxon OX14 4SB. Telephone: (44) 01235 827720, Fax: (44) 01235 400454. Lines are open 9.00–6.00, Monday to Saturday, with a 24-hour message answering service.

British Library Cataloguing in Publication Data
A catalogue record for this title is available from The British Library

ISBN 0 340 87262 4

First published 1999
Second edition 2001
Third edition 2003
Impression number   10 9 8 7 6 5 4 3 2
Year   2007 2006 2005 2004

Hodder Headline's policy is to use papers that are natural, renewable and recyclable products. They are made from wood grown in sustainable forests. The logging and manufacturing processes conform to the environmental regulations of the country of origin.

Cover photograph: Holly Harris/Getty Images
Typeset by Phoenix Photosetting, Chatham, Kent.
Printed in Great Britain for Hodder and Stoughton Educational, a division of Hodder Headline, 338 Euston Road, London NW1 3BH by Cox & Wyman Ltd, Reading, Berks.

Contents

Acknowledgements

First and foremost we would again like to thank our long suffering families for putting up with us during the time we spent thinking about, writing and putting together this handbook. Thanks must in addition be due to hundreds of our past students at Bury Grammar School, Adwick School and Winstanley College; their problems, perseverance and triumphs with coursework have been at the heart of our labours. To colleagues within our teaching institutions, who have provided help and advice, and to other colleagues in the examination world who have been generous with their time and provided us with extremely useful information on the practicalities of assessing coursework we also extend our thanks.

Malcolm Skinner, David Redfern, Geoff Farmer

Acknowledgements for the second edition

Geography at this level has moved on from Advanced to AS and A2, but the general principles that fashion and guide coursework remain the same. The real change has come with the emergence of new specifications and the different regulations that this has brought, and we have reflected this within the second edition of the book. Our thanks are once again to our families, colleagues and students for their patience, advice and, above all, commonsense.

M.S., D.R., G.F.

Acknowledgements for the third edition

The general principles of coursework do not change but many students are now taking the opportunity offered by the examination authorities to use their coursework in the context of an examination paper. In this third edition of the book, therefore, we have included a chapter that gives advice on how to approach this type of assessment. Our thanks are again directed towards our families, colleagues and students, but above all we must thank Alexia Chan and the team at Hodder & Stoughton for the vision that has directed the A-Z series into these new editions.

M.S., D.R., G.F.

The authors and publishers would like to thank the following for the use of illustrations in this book:

p19 (Figure 3.1), from *Fieldwork Investigations – Landforms* by Sue Warn; p36 (Figure 5.1) from *Rural Land Use and Settlement* by Sue Warn (1986). Both are reprinted by permission of Thomas Nelson & Sons Ltd.

A
B
C
D
E
F
G
H
I
J
K
L
M
N
O
P
Q
R
S
T
U
V
W
X
Y
Z

How to use this book

Geography coursework may have been around for a long time but this does not mean that students always find it easy to cope with the organisation of such work. It seemed to the authors that if Business Studies students could benefit from a coursework book, then geographers might similarly be happy to see a publication that would be as easy to use for coursework as the A–Z Handbook is for looking up geographical terms. So was born the *A–Z Geography Coursework Handbook* which has been written to be your personal guide to a successful project.

The bulk of the text of this book is organised by time. It starts where you, as a geography student, will have to start and works its way through 'title', 'collection', 'analysis' and 'writing up' in the same order as any student would normally follow. This structure should make it as easy as possible to use.

Will you as a student read this book from cover to cover? The answer to that question is probably that you will not; when the users of the book want to know how to develop a title then they will read the relevant section. When they want information on writing the report then they will read that section, and so on. The book will help you set up, research, write up and conclude an investigation. It is intended to serve as a permanent tutor/advisor, enabling you to see where to go next and ultimately how to maximise your marks.

The book falls essentially into three parts.

1   A step by step approach through a project from having an idea and developing a title through to the final writing up of the report.

2   A section containing information about the specific details of the requirements of each examination board and giving advice on how to maximise your mark in those situations.

3   A support section providing an A–Z of coursework terminology, resources and useful addresses.

Throughout the book there are four project checklists, each one providing an easy way of assessing your progress at various stages of the coursework process. Too much coursework is done at a rush towards the end of the specified period, and suffers accordingly when assessed. We hope that you will use the checklists to avoid this.

We have also provided you with a feature that should help you to put the advice given in this book into your own project context. From time to time there are sections entitled 'Projects in Practice' where we describe the experience of other students who have encountered similar situations to those detailed in the book. We hope that you will find their experience, and the ways in which they developed their coursework, helpful to your progress through a project.

Coursework has always provided students with a good opportunity to improve their overall grades. Written examinations can sometimes prove a bit of a lottery; coursework is very much within your control. You can spend a long time on your coursework but find yourself poorly rewarded unless your time and effort is directed towards the examination requirements. We hope that this book will keep your projects on the right track for success.

A
B
C
D
E
F
G
H
I
J
K
L
M
N
O
P
Q
R
S
T
U
V
W
X
Y
Z

Introduction

What is a geographical investigation?

All AS/A2 geography specifications require students to undertake investigative work. This work must be based on either primary data alone, or a combination of both primary and secondary data. In simple terms primary data is that which is collected by the student in the field, or material from other sources which has not been processed previously. Secondary data is information that has been derived from published documentary sources, and has been processed.

The requirement for primary research is paramount. Reading through textbooks, periodicals, newspapers and other journals is not enough. The writer must have had some direct contact with the subject of the investigation – for example, an actual area of study, a specific economic activity, or an identified group of people. A project may be based on an issue local to the student so that the writer is able to visit see, talk and listen.

It is necessary to state at this point that the following terms will be used throughout this book – investigation, enquiry, assignment, report and project. They all mean the same.

All geographical investigations should follow the same stages of enquiry:

* **the identification of the aim of the enquiry, often in the form of testing a hypothesis or establishing research questions, together with an indication of how it fits into a wider geographical context**

* **the collection of evidence or data, including measuring, mapping, observations, questionnaires, interviews and sampling techniques where applicable**

* **the organisation and presentation of the evidence using cartographic, graphic and tabular forms, possibly making appropriate use of information technology**

* **the analysis and evaluation of the evidence, noting any limitations of the evidence**

* **the drawing together of the evidence (synthesis), and formulating a conclusion**

* **the suggestion of possible extensions to the work, including further research questions which might be stimulated by the results**

* **an assessment of the success or otherwise of the enquiry, including an awareness of the significance of the work that has been completed.**

To an examiner, the investigation provides an opportunity to see what a student is capable of beyond the examination room. It provides an opportunity for the student to demonstrate initiative, perseverance, creativity and social skills. It is very important that you choose a project which will interest you – don't let someone else decide on this, although you should listen to advice. The more you are involved in the investigation, the more you will be part of it, and the more you will strive to excel in its completion.

Many A Level specifications state that there has to be a maximum number of words for the final report. Common word limits range between 3000 and 4000 words. Some specifications then state that reports which grossly exceed these maxima will be penalised in the presentation section of the assessment process. Consequently, it would be wise for students to find out precisely what the maximum limit is for their particular specification before starting work on the investigation.

Some specifications also require a particular format to the finished report, and again it would be wise to check this in advance. Features such as executive summaries, paragraphing, double spacing, and the use of bibliographies and appendices are examples of such requirements.

Finally, the proportion of the final marks that the investigation carries varies from specification to specification. It can vary from between 10% to 20%. Whatever the proportion, that is the amount of marks which is within your control. Follow the advice contained within this book, and all of your efforts will be channelled successfully into writing a better report of your investigation. Take advantage of this opportunity to boost your overall grade.

Projects in Practice

Colin lived in a semi-rural area on the outskirts of a town in northern England. The local council announced that it was examining the possible use of an abandoned quarry close to where he lived as a land-fill site for its refuse disposal. His parents became actively involved in the action group formed to resist the proposal. He decided that the issue could form the basis of his geographical investigation: he could write to the Council, interview local residents (including his parents), measure the possible environmental impact of the proposal and visit similar schemes. He was part of the issue – it was interesting, relevant and challenging. Perhaps his findings would be used in the decision-making process.

How to get started

1.1 Having an idea

Having an idea is only part of the process of establishing a focus for your investigation. The initial idea may arise out of one of the following sources:

* **the examination specification itself**
* **topics covered in lessons**
* **past titles of previous students in your school/college**
* **your teacher**
* **this book and other publications**
* **other areas of the media – television, newspapers etc.**
* **your interests, hobbies and other personal experiences.**

Having had an idea is only the start – more questions need to be asked before the details of the study can be finalised:

* **am I interested in what I propose to do?**
* **is what I propose geographical?**
* **will the information I need be available?**
* **is what I propose too simple, or too ambitious?**
* **is what I propose original?**
* **do I need to read about the topic?**
* **should I talk about my idea with others?**
* **where do I need to go to carry out the fieldwork?**
* **how easy is it for me to go to that area?**
* **is it safe for me to go there on my own?**
* **how much will it cost to get there?**
* **am I clear in my own mind what I intend to do?**

Aims and objectives

All these questions will help you to focus in on the precise aims of your investigation, and also assist in your methods for going about the task, the objectives. Aims should clearly state what you are hoping to achieve. Objectives should state how you will achieve your aims. You do not have to formally record these at the initial stage of your investigation, but having a short written note of some ideas or suggestions will help greatly as the investigation proceeds. Further guidance on both aims and objectives follows later.

Geographical topics cover a very wide subject area, ranging from physical geography topics to human geography topics to people-environment topics.

Topic: physical geography

Here is a list of possible subject areas in physical geography, which is by no means exhaustive:

* **a comparison of two contrasting coastlines, or two rivers, or two different parts of the same river**

* **a comparison of the climate of two contrasting areas – e.g. on opposite sides of a valley, between a rural and an urban area, between the edge and the middle of a wood**

* **a comparison of the soils of contrasting areas – along a catena, across different rock types**

* **a comparison of the vegetation of contrasting areas – e.g. across a sand-dune area, or an area of marsh, or through a woodland**

* **a study of change through time – e.g. of rivers, coastlines, vegetation and soils**

* **a study of distributions of geographical features – e.g. glacial features, drainage patterns, sediment characteristics, vegetation types, soil types and slope angles**

* **a study of processes operating in an area – e.g. weathering, mass movement, erosional, hydrology, depositional and climatic.**

Topic: human geography

Here is a list of possible subject areas in human geography, which is by no means exhaustive:

- a comparison of contrasting places – e.g. two or more shopping areas, two or more urban neighbourhood areas, the spheres of influence of two or more places

- a study of changes through time – e.g. of population characteristics of an area, the impact of social change on an area, the land-use changes within an area, the gentrification of an urban area, the changes in shop types within an area

- a study of people behaviour – e.g. shopping habits, movement to work, tourism and recreational activities, perceptions of people of distance or landscapes, the segregation of groups by age, income or ethnicity

- a study of distributions of human activities – e.g. schools, ethnic groups, medical facilities, old people's homes, crime, noise, disease and illness

- the testing of geographical theories – e.g. Christaller's Central Place theory, the theories of the internal structure of cities (Burgess, Hoyt, etc.), agricultural land-use patterns (Von Thunen).

Topic: the interaction between people and their environment

Here is a list of the possible subject areas in people/environment interaction; again by no means exhaustive:

- the study of local issues – e.g. changes to inner city areas, or outer city housing estates; the future impact of development of the Green Belt in an area; the provision for an increasing elderly population within a town; the impact or pattern of crime; the construction of a new road in an area; the impact of traffic calming in a residential area; the impact of a tourist development in an area; the impact of a new economic development on an area

- the study of the effects of people on the physical environment – e.g. footpath erosion; cliff instability; pollution patterns in rivers; flood control; dam or reservoir construction; afforestation or deforestation; pollution patterns in the atmosphere.

A B C D E F G H I J K L M N O P Q R S T U V W X Y Z

Projects in Practice

Natalie lives in a village on the outskirts of Reading. Near where she lives, the owner of an 11 acre area of land, which is mostly in disrepair, has recently died, bequeathing the land to his niece and her family. Before his death, the owner was repeatedly approached by developers offering to purchase the land, which he turned down. In his will he stated that the land should not be redeveloped without the consent of the two immediate neighbours. Although this is not a legal requirement, the niece would like to respect these wishes.

Natalie believes that there is the potential here to undertake an enquiry into what would be the most suitable form of redevelopment – housing, agricultural or recreational? The area of enquiry is literally on her doorstep.

1.2 Preparation and planning

Successful fieldwork depends on careful planning. Such planning should begin as early as possible in the A Level course, probably in the first term, and certainly by the end of the first year. Early planning enables you to have plenty of time to think of a suitable topic for study, to discuss it with your teacher/tutor, and possibly to have some feedback from an Examination Board Moderator if that is a feature of your specification. Some specifications stipulate the criteria by which the study proposal will be judged by a Moderator and it would be wise to check on these at an early stage. Other specifications allow a free choice of topic.

Many students choose topics which are based on their local area, but you are not necessarily bound by this. You could choose a topic for study which is located within a day's journey from your home. More distant locations are possible, but you should be fully aware of the difficulties created by inadequate and incomplete data collected in distant locations. It may be difficult to return to the area to add to the data should it prove to be necessary. Most specifications allow students to collect data as a group, indeed for many topics it is safer to do so. However, it is also often the case that each student should have a unique title, and you need to check that your friends/colleagues do not have the same idea as you.

Types of fieldwork

The term 'fieldwork' covers a variety of activities, and it is important at this stage to describe each of these activities as you may wish to include one or all of them in your work.

1 The guided tour

You are taken around an area by an 'expert', often your teacher, and told all about it. The important point about this type of fieldwork is to research the area in question before you visit it. The more you know about the background of the area, the more you are likely to see.

2 The experimental piece of fieldwork

You are trying to find out answers to previously unresearched problems. The result of the research is not known before the work begins. This type of fieldwork is often used in examining local issues and conflicts, where more experienced researchers have not yet had the time or the inclination to investigate the issue. This is an opportunity to break new ground, and is more interesting for you as a result.

3 The 'pseudo-experimental' piece of fieldwork

Data is collected, but the outcome of the investigation is probably already known. This type of fieldwork is similar to an experiment in a science lesson – you are simply providing evidence to support a theory. A great deal of geographical fieldwork is of this nature: for example to measure variations in the characteristics of a river as it flows downstream, or to measure the extent of a Central Business District within a town. This is not original research, but it does not have to be. Some of the conclusions may tell you things you did not already know, while others will confirm what you did know. At the very least, the processes of data collection, the presentation and analysis of your results, and the formulation of conclusions are worthwhile exercises in their own right.

Which type of fieldwork you choose to adopt, or combine, will depend very much on the nature of your investigation. However, it is clear that you must choose what you can manage in terms of both time and cost, and what is most appropriate. Do not do a certain type of fieldwork for the sake of doing it – you can make significant mistakes by doing so. This all forms part of the planning and preparation stages of your investigation.

Projects in Practice

Julia lives in an area in the East of London. She has relatives who live on the Isle of Wight and she visits them during the summer holidays. Last summer she noted that the main coastal road which connects one part of Freshwater Bay with another part of the island was under threat from sea erosion and landslips. The local council was examining the possibility of re-routing the road across an area of protected chalk grassland. Julia decided that this could form the basis for an investigation – it was a local issue which had economic, social and environmental impacts. However, after collecting some data during that summer, the road collapsed into the sea during the following winter. Julia was unable to complete the enquiry, or to bring it to a satisfactory conclusion as she lived too far away from the area. What had begun as a good idea became unmanageable in the time allowed.

How to develop a title for your investigation

The best type of investigation involves the testing of a hypothesis, or the setting of a research question, that can be investigated and evaluated. A hypothesis is a statement that is made about a geographical question. The question may be 'Do similar land-uses cluster together in a town centre?' and the hypothesis could then state, for example, 'There is a significant concentration of shoe shops in one part of the shopping centre'. The statement, or hypothesis, can now be tested by the collection of data which is easy to identify – the location of shoe shops in the shopping centre of a town. An answer to the original question will also now be capable of being given with reference to shoe shops.

The following illustrates the difference between each of these approaches. A bad project title would be:

'A study of the old people in Doncaster'. Here, there is no question and no hypothesis.

A better title, involving a research question, would be:

'What factors determine the distribution of old people within Doncaster?' This establishes a question to which an answer can be developed.

An alternative title, in the form of a hypothesis, would be:

'The distribution of facilities for old people in Doncaster does not match the demand for their health care'.

Either of these latter two approaches are acceptable ways in which to proceed.

2.1 What are the background sources which you use for your investigation?

These are concerned with the preliminary material and work which preceded the study; in a sense what 'triggered' the investigation in the first place. Your study may have developed from a desire to test out some aspect of a theory introduced in class, or in a textbook, or an idea stimulated by a newspaper article or letters relating to a conflict in your local area, or an awareness of a general problem such as the impact of tourism in an area of high landscape quality. You may have seen a video about tourist pressure in honeypot sites and its effect on footpath erosion. This may then stimulate the hypothesis 'footpath erosion decreases with distance from a car park'.

2.2 What are the aims of your investigation, and what research questions can be developed?

You need to avoid being vague about the purpose of your study: 'The aim was to study a beach... a river... shopping habits... soil types'. No real purpose is identified by these statements. You need to be more precise in what you are seeking to do:

* **'my aim is to study changes in vegetation types in a sand-dune complex with increasing distance from the high water mark'**

* **'my aim is to investigate the relationship between soil characteristics and slope angle along a transect across a valley'**

* **'my aim is to investigate the pressure on St Albans' transport system, and to study the impact of the proposed bus lane system in the town'.**

Research questions then clearly develop from these aims:

* **'does the number of vegetation species increase with distance from the sea?'**

* **'does soil acidity change downslope?'**

* **'where will the new bus lanes be located, and what times of day will they operate within?'**

Projects in Practice

Michael lives in an area near Wimbledon Common in London. He regularly walks his dog across the Common. He thought that it would be good idea to base his enquiry on the Common, and that his aim would be to examine the impact of recreational activities on the environment of the Common. But how could this translate into research questions? He came up with the following:

Continued

- What are the recreational activities on the Common?
- What is the ecological nature of the Common?
- What effect does each recreational activity have on the ecology of the Common?
- What management strategies currently exist within the Common?
- In what ways could these strategies be improved?

These research questions provided a logical route through the enquiry, and Michael knew clearly what he had to do.

What is a null hypothesis?

A more sophisticated way to establish a hypothesis is to set it up in two parts: a 'null hypothesis' and an 'alternative hypothesis'. The former takes the form of a negative assertion which states that there is no relationship between any two variables which are being tested. For example, a null hypothesis could state that 'there is no relationship between building height and distance from the centre of a central business district (CBD) of a town'.

This assumes that there is high probability that any observed links between the two sets of data are due to unpredictable factors. If building heights were seen to decrease with distance from the CBD in the chosen town, then it was as a result of chance, or luck. However, if the null hypothesis can be rejected statistically, then we can assume that the alternative hypothesis 'that there is a relationship between building height and distance from the centre of a central business district of a town' can be accepted.

One benefit of this seemingly 'reverse' approach to an investigation is that if the null hypothesis cannot be rejected then it does not mean that a relationship does not exist – it may simply mean that not enough data has been collected to reject it. This does not mean that the exercise was worthless – it was just too limited in its scope.

Another benefit of this approach is that it allows the use of statistical tests on the significance of your results to be carried out. More details on these tests appear later in the book.

Examples of titles that are potentially successful

1 What factors influence the rates of gravestone weathering?

2 How does longshore drift vary along a stretch of coastline?

3 What factors affect stream velocity?

4 How does the load of River A differ from that of River B?

5 What impact will the new Selby by-pass have?

6 The environmental impact of the widening of the A33 around Eastleigh will be greater than anticipated.

7 What are the tourist pressures facing Hay Tor?

8 Is recycling a viable option in woodlands?

9 How effective are the coastal defences at Withernsea?

10 Stillingfleet mine in the Selby coalfield has had a very different impact when compared with the impact of Askern pit in South Yorkshire.

11 What impact does rock type have on the characteristics of soils?

12 How do soils vary with slope?

13 Afforestation in the Kielder area has had significant social and economic effects.

14 The design of housing estates has an impact on the patterns of litter, vandalism and graffiti within them.

15 Does York offer a full range of tourist developments that make full use of its historical legacy?

16 The demographic transition model can be applied to the population growth of a village in Southern England over the last 100 years.

17 The degree of flood control in the Maidenhead area is disproportionate to the flood risk in that area.

18 What are the factors influencing industrial location decisions in the Swindon area?

19 The distribution of crime in Wolverhampton is related to housing deprivation.

20 The distances between premarital addresses of marital partners increase over time.

21 Vegetation patterns in a salt-marsh are related to tidal factors.

22 What are the changes within the psammosere succession at Braunton Burrows, Devon?

23 How do flow patterns vary within the channel of the River Bollin?

24 Tourism has caused significant conflicts in the Castleton area of Derbyshire.

A
B
C
D
E
F
G
H
I
J
K
L
M
N
O
P
Q
R
S
T
U
V
W
X
Y
Z

How to measure the feasibility of your investigation

Having decided on the broad area of your investigation, and having decided on the nature of the fieldwork to be completed, you then need to check whether or not the investigation can proceed to a satisfactory conclusion.

Here are some questions which you can ask yourself to see if your idea can really work.

3.1 Is my topic geographical?

You need to check that your project deals with the characteristics of places or features, or examines distributions and patterns, or looks at the relationships between people and the environment in some way. If so, then the topic is geographical, and you should proceed.

3.2 Will I be able to collect and analyse suitable data?

Unfortunately, many potential investigation titles are not suitable because there is very little data which can actually be collected about the topic.

There are two types of data which can be collected. Primary data is that which you collect yourself (first-hand), and secondary data is that which has already been produced and published by someone else. Each of these types can further be subdivided into quantitative data and qualitative data. The table opposite gives some examples of each of these types of data.

	Primary	Secondary
Quantitative	Land-use transects	Census data
	House type survey	Marriage registers
	Environmental impact assessment	Traffic flow data
	Traffic counts	Meteorological data
	Noise/pollution surveys	Government statistics
	Climate surveys	Directories
	River measurements	
Qualitative	Questionnaires	Newspaper cuttings
	Interviews	Leaflets and brochures
	Field sketches	TV/radio programmes
	Photographs (yours)	Company publications
		Photographs (not yours)

All of these types of data are normally used in investigations. All examination boards place emphasis on the data-collection element of the investigation. For this reason it is wise to come up with an idea which enables you to collect and use a wide range of data sources, and not just one. Don't just base your investigation around questionnaires.

Projects in Practice

Eve decided to carry out an investigation into the development of coastal defences between two small seaside towns of the south coast of England. She knew that she needed to make use of a range of both primary and secondary data sources.

The primary sources she used were:

photographs, field sketches, land-use mapping, measurement of beach profiles, questionnaires, interviews, an ecological survey and an environmental impact analysis.

For secondary data, she obtained information from:

the local council, textbooks, newspaper articles, leaflets, photographs in the local museum, an Ordnance Survey map, and the research of a local environmental group.

What makes a good questionnaire?

Questionnaires and interviews are the best methods of obtaining up-to-date information about people and their habits. You should bear the following points in mind:

* **interview enough people – 50 is often a good number as it means that you are likely to obtain a good sample or cross-section of people and views**

* **be polite and explain why you want to ask people questions – do not be upset if some people refuse to help you**

* **keep the questionnaire short as most people are busy – four or five questions are usually sufficient for a questionnaire**

* **ask clear questions which are not too complicated – do not give too many alternative responses**

* **when asking for information that may be personal, such as age or income, then give people a category into which they can fit without being so precise that they become concerned – for example a person may admit to being between 18 and 35 years old, but would not want to admit to being a particular age**

* **wear smart clothes – and be ready with a smile and a thank you!**

3.6 What equipment will I need?

The amount and nature of the equipment required is obviously an element of project feasibility and will obviously depend on the nature of the fieldwork being conducted. Here are some suggestions that may apply:

* **a clipboard, paper (plenty of it) and a pen/pencil/ruler**

* **a plastic cover for the clipboard in case of rain**

* **a map of the area being examined**

* **binoculars are useful if you need to look at far away places, especially where access is not easy**

* **a compass if direction is an important element of the investigation**

* **a camera is often useful to record either a feature you are studying, or to record you and your friends carrying out the fieldwork**

3.3 Is the subject matter narrow enough?

In general, it is better to study one aspect in detail than several aspects sketchily, to study one river thoroughly rather than five in brief.

The size of the area to be studied will depend on the topic chosen. Studying settlement hierarchies means that you need to visit an area large enough to contain a significant number of settlements of a variety of sizes. On the other hand, a study of social change or social behaviour within an urban area may well be conducted at a scale of a few streets, especially if detailed census data is available. A project based on microclimates will be more successful if based on the climate around a few buildings, than a more generalised study across a whole city.

Sample size is a major consideration here. It is important that the area studied should be representative enough to enable you to draw conclusions which can be meaningful. It is unsound to draw general conclusions from a small sample, say one farm or one factory.

3.4 Will I be able to complete the investigation within the time period allowed?

One type of investigation which needs to be treated with caution is that of the study of the impact of an event, say the impact of a new by-pass around a town. This type of project can be very successful, but suffers from the timescale within which you, the student, have to work. It is usually impossible to collect data both before and after the event has taken place. To measure the impact of a new by-pass around a town is very difficult without both 'before' and 'after' data.

3.5 Will a questionnaire be of use, and how shall I carry it out?

The use of a questionnaire is a popular way of collecting primary data. It is important though that you prepare and test your questionnaire before you begin the investigation proper. Try it out on your friends and family to see if it gets the response you want or need.

As a precaution it is always better to seek approval from your teacher or tutor before you start. This will avoid insensitive questions, and prevent harassment of local people by too many questionnaires. When carrying out a questionnaire, NEVER work alone.

- soil studies may require a spade or trowel, an auger, a soil acidity testing kit, and polythene bags for collecting samples together with clips

- pollution studies may require a thermometer, nets and trays, pH papers, and bottles (with tops) for samples

- vegetation studies may require a quadrat, tape measures, reference books to help recognise plants, a clinometer for measuring slope angles, polythene bags for samples or sellotape for attaching small samples to paper

- climatic and weather studies may require equipment such as thermometers, an anemometer, a wind vane, a rain gauge and a barometer

- studies involving river work can use a variety of different pieces of equipment, for instance tape measures, ranging poles, floats of various kinds and a stopwatch for measuring the speed of the river or a flowmeter, a clinometer to measure the gradient of the stream, a stone-board for measuring the length, shape and roundness of stones on the bed of the river

- studies involving coastal work can also use a variety of different pieces of equipment, for instance tape measures, ranging poles, a quadrat, a clinometer to measure the gradient of a cliff or wave-cut platform, a pantometer for measuring the slope of a beach, a stone-board for measuring the length, shape and roundness of pebbles.

- finally, do you need to be reminded to take sufficient food and drink?

Projects in Practice

As part of her enquiry, Michelle investigated the impact of mountain biking on the footpaths of an area. She wanted to measure or assess the amount of erosion that had been caused, and the amount of compaction of the ground that had occurred. To do this, she needed a tape measure, a metre ruler, a smaller ruler, a tin can opened at both ends, some water, and a stopwatch. She also decided to photograph many of the sites, and so needed to borrow a camera and to buy a film.

3.7 Is there anything else that may be able to help me?

Here are a number of suggestions that might make your study appear more feasible:

A number of items may fit into this category.

1 **Would any of your other subjects help you? For instance if you are studying chemistry you might be interested in examining the chemical composition of different horizons in soil profiles. Or, a study of pollution in a river may be worth consideration. If you are studying biology, then the recognition of flora and fauna changes in an area may be easier for you.**

2 **Do you have any special interests or hobbies which might be relevant? If you are a keen angler, you may be interested in investigating stream pollution.**

3 **Do you have any special contacts or sources of information? You may have access to the records of a family business which might be relevant to a study of the changing location of markets for a particular product.**

Projects in Practice

Mark is a keen fisherman. He is also interested in the biological aspects of the local brook where he fishes. He decided to carry out an investigation into the extent of pollution of that brook. He wanted to measure the following aspects, using the identified equipment:

- oxygen levels – an oxygen meter
- nitrate levels – nitrate strips (which also measure nitrite levels)
- acidity levels – pH indicator strips
- species types – a 'D' frame net.

3.8 Is there anything else I need to do?

1 **Always ask permission if it is needed**
Any investigation that involves you going on to someone else's land, or building, will require permission from the owner of that land or building. It is important therefore that you get permission to enter and to work. In most cases, a telephone call or a letter explaining the purpose of your visit will suffice. It is better if you can also ask your teacher or tutor to write a short note 'To whom it may concern' on school or college headed notepaper in support of your work. However, remember that they are busy people too, and need to be given time to produce the letter well before you need it.

2 **Check if you can get there, and back**
Another consideration is transport to and from the area(s) where you are studying. How accessible are the places? Will you need help to get you there and get back?

3 **Wear appropriate clothing**
You need to wear suitable clothing for outdoor work. If you are going to remote and difficult areas then you must wear clothing that will keep you warm and dry. Wearing appropriate and strong footwear, even when working in urban areas, is a good idea. Wellington boots would seem to be a must when working in or near a river.

4 **Be safety conscious**
It has already been stressed that you should not work alone. Always let someone else, preferably an adult, know where you are working, when you will get to the area of study and when you intend to return to your base or home. If working in mountainous areas, it is better to complete the work in summer when the weather is better. When working in coastal areas, always check on the tide levels both before and as you are working. Never wade around headlands unless the tide is clearly going out. Never stand or work underneath crumbling cliffs, and do not collect samples from cliff faces. Always stay on coastal paths.

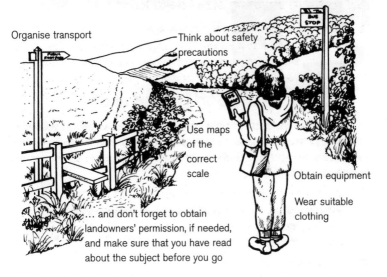

Organise transport

Think about safety precautions

Use maps of the correct scale

Obtain equipment

Wear suitable clothing

... and don't forget to obtain landowners' permission, if needed, and make sure that you have read about the subject before you go

Figure 3.1 Ready to go?

A B C D E F G H I J K L M N O P Q R S T U V W X Y Z

Setting the scene

The first part of your report should introduce the area or topic of your study. This will enable you to explain your reasons for choosing the topic, and allow the reader to become familiar with the geographical context of the investigation.

4.1 Introducing and explaining your investigation

It is often the case that a proportion of the marks to be awarded for the investigation as a whole is for 'Setting the scene'. This introductory section should:

- **introduce the broad area of concern**
- **explain the reasons for study**
- **refer to the key questions that are being asked or the hypothesis being tested**
- **provide sufficient general background (including location maps) necessary to introduce the topic**
- **mention any possible wider significance or broader implications of the study.**

The following examples taken from A Level projects may be helpful in seeing how successful students have been able to set the scene for their investigations.

Projects in Practice 1

Title: How effective will the proposed flood control scheme in XXX be?

The towns of North Yorkshire such as York and Selby ▶ have an extensive history of flooding from the River Ouse

---

and its tributaries. Documentary evidence of such flooding dates back several centuries, and flooding has taken place regularly in recent times.

The reasons for the flood risk are both physical and human. The Ouse has a number of tributary rivers (Swale, Nidd, Wharfe, Aire) which drain into it, each of which flow from the Pennine hills. Rainfall totals are higher in these upland areas, and there are periods of extensive snowfall at regular intervals which cause problems when temperatures rise. In addition, urban growth is continuing to take place within the Vale of York at towns such as Harrogate, Boroughbridge and York itself. These new urban areas increase the amount of impermeable surfaces and underground drainage systems all of which speed up the time taken for rainwater to reach the rivers.

The National Rivers Authority (NRA) continue to monitor the flood risk, and to research into possible ways to mitigate the flood risk. A number of schemes have been set up within the York urban area itself, as well as wider flood alleviation schemes in other parts of the drainage basin. Of the numerous options available, a scheme involving the construction of an alleviation channel around the town of XXX, together with alterations to the existing channel has been suggested.

The aim of this study is to assess the impact of the proposed flood alleviation scheme. The complete scheme is too large to investigate in detail hence I have decided to concentrate my enquiry on two sections of it. The two sections I have chosen are identified on the location map provided.

The following research questions will be applied to each site:

- what is the present situation in terms of land uses?
- what does the proposed scheme involve?
- what will be the social, economic and environmental impacts?
- how will the interests of various people and groups conflict with the scheme?

Continued

They will be answered using a combination of primary and secondary data sources and variety of presentation techniques. The answers to the research questions will be compiled in the conclusion at the end of the study and compared with other flood alleviation strategies employed by the NRA. Through a comparison of this strategy with other case studies it may be possible to ascertain whether some flood alleviation schemes are more successful than others, and whether some are more environmentally sensitive than others.

Projects in Practice 2

Title: Conflicts have been caused by tourist development in XXX.

XXX, population 689 (1991 census data) is situated in the Peak District National Park. The Peak Park is the most heavily visited UK national park attracting around 22 million visits per annum. This is mainly due to its proximity to the large urban conurbations of Sheffield, Leeds, Manchester and Nottingham. It is estimated that half of the population of England live within 100 miles from the limits of the Peak National Park.

XXX is one of the most popular tourist centres in the Park, and lies within the Hope Valley. It attracts approximately 2.25 million visits per year. Tourism in this area has increased for a number of reasons:

- improved transport has increased access and mobility
- more tourist accommodation has been constructed
- people have greater disposable incomes
- people have more leisure time, including paid holidays
- people have shorter working hours, increasing leisure time
- more people have longer periods of retirement, and so have more leisure time.

Continued

XXX is a tourist honeypot. The village and surrounding area have many attractions for visitors such as caves, gift shops, and unspoilt scenery. The area has a high landscape value due to its vegetation, geological and historical aspects. The area also has a number of sites of special and scientific interest (SSSIs).

The Peak Planning Board has designated XXX a Conservation Area. This shows a clear recognition of the special historical and architectural character of the area. A number of enhancement works have followed including paving, tree planting, the removal of overhead wires and the renovation of the village square.

A large proportion of employment is generated by tourism in the village. However, increased visitor numbers are leaving a marked impact on the built, social and physical environments. This is creating conflict within the local community, between those benefiting from tourism and those inconvenienced by it.

The objective of this investigation is to identify the likely positive and negative effects that tourism has had on XXX. The impact of tourism on each of the following will be examined:

- the environmental quality of the area
- the local economy of the area
- the provision of facilities within the area
- traffic flow and parking.

The report will also investigate possible schemes and solutions to problems caused by increasing tourist numbers.

Projects in Practice 3

Title: Variations in velocity patterns, bedload and sinuosity existing along the channel of the River ZZZ.

The River ZZZ is a tributary of the River Cherwell in Oxfordshire. This project proposes to investigate flow patterns within the channel of the river at five sites along its course. Linked to this I shall also be examining the bedload content of the river at each of these five sites.

The following research questions summarise the main features of my investigation:

- do velocity patterns change both in cross-section and in magnitude downstream?
- does bedload content change downstream?
- does sinuosity increase or decrease downstream?

I assembled a number of items of equipment for the fieldwork that I undertook. These included tape measures, a flowmeter, a metre rule, two ranging poles, a stopwatch, a sediment trap, soil sieves and local scale maps.

The five sites that I visited were easily accessible, and were spread out as evenly as possible along the river's course. The total length of the section that I investigated was approximately 11 miles. The fieldwork was conducted mostly during the summer months of my Lower Sixth year. I revisited one of the sites in the October half-term to complete another flowmeter test as I had lost the results from the previous visit.

4.2 The use of location maps

Location maps are a useful means by which general background material can either be provided or enhanced. The nature of the map or maps will depend on the issue being investigated. However, some general considerations are as follows:

- **where possible hand-draw the map and do not photocopy it**

- **if the map is to be so complex that you cannot hand-draw it, then one option could be to use transparent overlays for labelling and annotation**

- **the use of a number of maps at a variety of scales could be worthwhile – for example a local large scale map to show the immediate area, together with a smaller-scale map to show the regional or national context of the issue.**

Figure 4.1 (page 26) shows an example of local scale map introducing the conflict surrounding the Newbury by-pass. Figure 4.2 (page 27) illustrates the national context of proposed new housing developments in the south of England.

4.3 The importance of using checklists

When examiners comment on investigations there are a number of criticisms which are common to the work of all students whatever course they are taking. Some of the main criticisms have been that the authors are not always clear about their aims and have put far too little work into field research, resulting in the collection of nowhere near enough data for analysis. For a number, the date upon which they began their fieldwork was too close to the submission date for their work leaving very little time for analysis and evaluation. For some there was a weak geographical link or the scale of the study was inappropriate for the proposal.

It is important that all candidates should guard against such weaknesses. It has been the opinion of senior examiners for a long time that candidates should check their work rigorously at each stage of its development. To this end we have produced checklists, which you will find throughout this book, placed to coincide with crucial stages in the production of the finished piece of work. Read each one and if you cannot answer yes to each point checked, then consult the relevant sections of this book or your supervising teacher or even both.

Remember that the examination boards have slightly different procedures for investigations, so although we have tried to write these lists to cover all possibilities it may well be that certain questions do not apply to your particular course, although we are certain that almost all of them will at some stage.

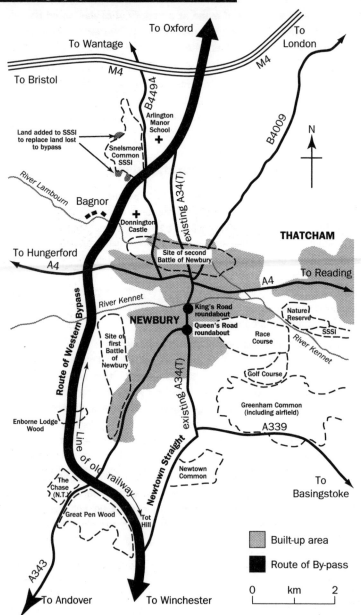

Figure 4.1 The Newbury by-pass

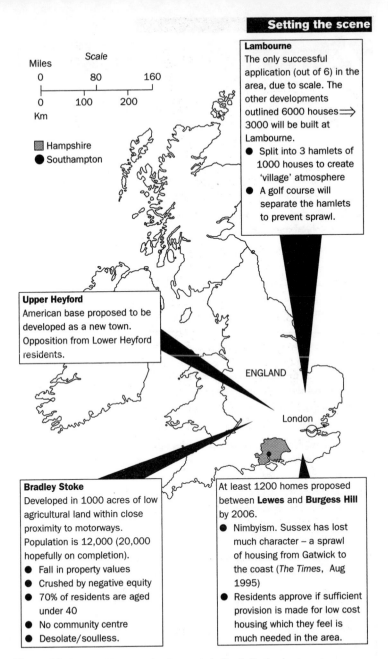

Scale

Miles
0   80   160

0   100   200
Km

▨ Hampshire
● Southampton

Lambourne
The only successful application (out of 6) in the area, due to scale. The other developments outlined 6000 houses ⟹ 3000 will be built at Lambourne.
- Split into 3 hamlets of 1000 houses to create 'village' atmosphere
- A golf course will separate the hamlets to prevent sprawl.

Upper Heyford
American base proposed to be developed as a new town. Opposition from Lower Heyford residents.

ENGLAND

London

Bradley Stoke
Developed in 1000 acres of low agricultural land within close proximity to motorways. Population is 12,000 (20,000 hopefully on completion).
- Fall in property values
- Crushed by negative equity
- 70% of residents are aged under 40
- No community centre
- Desolate/soulless.

At least 1200 homes proposed between **Lewes** and **Burgess Hill** by 2006.
- Nimbyism. Sussex has lost much character – a sprawl of housing from Gatwick to the coast (*The Times*, Aug 1995)
- Residents approve if sufficient provision is made for low cost housing which they feel is much needed in the area.

Figure 4.2 Proposed housing developments in South England

✓ Checklist 1A

Is my topic suitable?

These are the questions you should be asking early in the process when you have developed a proposal for submission.

	Yes	Partly	No
1 Have I read the specification document to ensure that my proposal will meet all the requirements?	☐	☐	☐
2 Does the topic that I have proposed have a strong geographical component?	☐	☐	☐
3 Is the scale of the topic appropriate?	☐	☐	☐
4 Is it possible to state the core of the topic in a few words, as the title of the investigation, either in the form of a question or a relationship to be examined?	☐	☐	☐
5 Does the proposed topic relate to some part of the specification that I am studying and can I establish that link in the study backed up with appropriate references to the relevant reading?	☐	☐	☐
6 Will the topic allow me to collect sufficient primary data or evidence as a basis for analysis?	☐	☐	☐
7 Do I have the time and expertise to collect the amount and type of data needed to produce worthwhile and reliable results?	☐	☐	☐
8 Have I built sufficient flexibility into the topic to allow me to modify, or narrow, the focus at an early stage so as to meet the investigations's revised aim and make best use of my time in the field?	☐	☐	☐
9 Will I have access to the equipment and transport required for the whole period of data collection?	☐	☐	☐

✔ **Continued**

	Yes	Partly	No
10 Can I gain access to the field site and any restricted data sources such as parish records and if permission is required, has it been obtained?	☐	☐	☐
11 If interviews form part of my primary collection, have I checked that the people concerned will see me and provide the information that I need?	☐	☐	☐
12 Is the data and other information that I am planning to collect capable of being analysed and presented using appropriate techniques?	☐	☐	☐

A B C D E F G H I J K L M N O P Q R S T U V W X Y Z

✓ Checklist 1B

Am I prepared for fieldwork and data collection?

You should ask these questions after the approval (or modification) of your proposal and before beginning the fieldwork and data collection process.

	Yes	Partly	No
1 Having received the comments from the moderator/adviser, do I understand what is being said and have these observations been discussed with my supervising teacher?	☐	☐	☐
2 Do I have the timetable drawn up to ensure that I complete each stage of the fieldwork in time to allow for analysis of the data?	☐	☐	☐
3 Have I read and noted sources of background material?	☐	☐	☐
4 Do I know which techniques I am going to use in the analysis of my field data and have I allowed for any requirements of those techniques which may influence my collection of that data (sample size, for example)?	☐	☐	☐
5 Have I obtained the correct maps for my investigation and have the base maps that I may need in the field been drawn up?	☐	☐	☐
6 Do I have sufficient copies of my data recording sheets and have I had a trial run to ensure that they will allow me to record the data that I need?	☐	☐	☐
7 Have I piloted my questionnaires and if necessary made alterations? Do I know (a) the size of the sample I need and (b) the ways of ensuring that this sample is representative?	☐	☐	☐

Collecting the data: primary research

You will have already begun to think of the data that you will require so that your investigation can proceed to a satisfactory conclusion. The examination boards insist that the data used for such investigations has in part to come from your own personal observations, i.e. must involve the collection of primary data through such techniques as questionnaires, interviews, transects, river measurements, traffic and pedestrian surveys, etc. You can base your investigation wholly on such data, but in general, and depending upon the work in question, it should also be possible to include some secondary material.

5.1 Sampling

It is usually neither possible nor necessary to collect enormous amounts of data. The idea behind sampling is that you will be able to obtain a representative view of a geographical feature, issue or problem by collecting small amounts of information which are carefully selected. You cannot for example look at all the pebbles on a beach or seek information and opinions from all the shoppers visiting a market centre on the same day, but you can look at a fraction of such populations in the context of your data collection.

Once you have established that you will need to carry out some form of sample you need to choose a method which will ensure that data is collected in sufficient quantity and in an objective way. If, for example, you wish to interview shoppers, and it is clear that the target population consists of a range of ages, there is no point whatsoever in choosing to interview only the young attractive respondents. There are a number of sampling techniques, but the ones which geographers use most frequently are random, systematic and stratified (or quota).

5.2 Random sampling

As the name implies this form of sampling is generated randomly often by using random number tables (see the random number table on page 42 in section 5.9 Point sampling). A random sample is one that shows no bias and in which every member of the population has an equal chance of being interviewed or used.

Projects in Practice

David was seeking within his investigation to establish the market area of a number of small market towns in a mainly rural county in England. Part of the investigation would be to ask villagers where they shopped for various goods and thus establish patterns of movement. In each village David decided that be would interview at least twenty respondents but in a random way. Using a map of the village he numbered each house and then used random number tables to select addresses for his interviews.

5.3 Systematic sampling

In this case the sample is collected in a consistent manner by selecting, for example, every tenth person or house. The process is repeated until the sample total is reached. In a village survey, you could walk through the settlement and try to speak to the inhabitants of every tenth house in order to obtain a 'ten per cent sample'. On a beach you could record data on pebbles every 50 or 100 metres along the area.

5.4 Stratified sampling

This form of sampling requires that you know something in advance about the area in question or your target population. For example, if you know the age profile of people in an area then the sample must reflect that age distribution. In such a case you should set the sample quotas as follows:

Demographics	Profile of area	Sample size (total 50)
15–30 years	36%	18
31–60 years	40%	20
over 60 years	24%	12

In area sampling, with a land use survey for example, if you know the soil of an area then sites should be taken in proportion to the area covered by each type of soil. A soil that covers 50 per cent of the area in question should therefore have half of the total sample points taken within its area.

5.5 Sample size

Once you have established a method of sampling it is time to consider the sample size that you will require. This will often depend upon the complexity of the survey used. With a questionnaire, for example, where a number of questions are involved it is necessary that the number of respondents questioned needs to be large enough to take account of the considerable variety introduced by the range of questions used. But, what in an investigation should be the actual sample size? In the process of sampling one has to accept a trade-off between the practical difficulties that limit the size and selection of the sample, and the reliability of the results. Within the limits of efficient sampling it should therefore be the aim to keep the sampling error as small as possible. Sampling error is generally dependent on three major factors:

* **the way that the sample is selected or designed**
* **the size of the sample**
* **the variability within the population being sampled.**

In a projects of this kind is largely impractical to expect you to carry out sample surveys to the same degree as professionals, i.e. running into several hundreds, in order to ensure that the sampling error is at a minimum. On the other hand, it is clear that in surveys such as shoppers in a market centre, then a sample of 20 to 30 would be so unrepresentative of a population as to be of little use.

As a guideline:

* **if a questionnaire is used for sampling then the sample size should reflect the number of questions asked**

- **sample sizes of 20 to 30 are often far too small, leading to unrepresentative results, particularly on surveys dealing with catchment areas**

- **the sample size should give all members of the population it represents an equal chance of being included.**

Projects in Practice

Penny was conducting an investigation in which she had decided to undertake a shopper survey using a questionnaire. Several questions would have to be included such as reason for visit, place of origin, frequency of visit, transport used, etc., in order to seek relationships between various aspects of the respondents' behaviour and to help account for the catchment patterns revealed. She realised that the sample size needed to be large enough to take account of the considerable variability which could be introduced into the sample by the range of questions used. Penny finally decided that she needed to interview at least 100 people and that to avoid bias she should carry out the survey over a number of different days.

5.6 Bias

Bias occurs in sampling when there is some distortion or error in the sampled data and the sample therefore becomes unrepresentative of the population in question. Bias usually results from a poor choice of method or when an insufficient number has been selected. In a shopper survey, for example, it is not a good idea to conduct the interviews on the same day of the week. It is sometimes possible, if it is recognised from the outset, to turn an element of bias into an advantage. In a beach survey on longshore drift, for example, you could select pebbles of one particular rock type (if you are able to recognise it!). This would have the advantage of showing how one type of rock was moved where rock patterns could actually be quite complex and difficult to interpret due to the variety of rock types involved.

5.7 Questionnaires

For many AS/A2 Level investigations one of the major methods of sampling is using a questionnaire. Writing a questionnaire is often time consuming as it is one of the most difficult aspects of individual investigations. Many examiners at this level comment upon the fact that questionnaires are often badly designed, with an insufficient balance between the more specific questions, which can provide objective data for mapping and quantification, and the more open-ended types of question (what is your opinion of …?). Such questions may be less quantifiable but can provide additional information that helps to explain the respondent's behaviour as revealed in the answers to the more factual questions. Too often candidates put a lot of effort into a questionnaire survey, shoppers at a supermarket for example, only to find that the survey reveals no more than the most basic information on the spatial and temporal behaviour of the shoppers. Where there were oddities in the patterns revealed, these could not be accounted for because the candidate included no open-ended questions to get at shopper's motives for their decisions.

It is possible to put even the more open-ended questions in a form that makes answers to them quantifiable. For example, for a question that begins 'would you say that your opinion on…' you could put in a choice of boxes to tick for the respondent's answer such as strongly in favour, in favour, undecided, against, strongly against, etc.

Many questionnaires can become far too sociological, and therefore somewhat limited geographically, by not concentrating on the spatial aspects of the sample being studied. Such surveys become too concerned with age, gender, attitudes, etc., and therefore give little scope for mapping and analysis of pattern. Be careful that your investigation does not become a social survey showing, for example, how shopping frequency is affected by age or gender with almost no reference to the distance and spatial elements.

It is important that studies of this kind at AS/A2 Level should be very different from the work submitted at GCSE. At GCSE it may be acceptable to undertake a small-scale questionnaire simply in order to demonstrate your ability to do it. This is not sufficient at AS/A2 level where the quality, quantity and the reliability of the data produced is as important as the means by which it was obtained.

Questionnaires include different types of questions:

> **Closed questions**: these give a choice and so limit the freedom of response – they are quicker to deal with and produce quantifiable results. Forms of question can take a yes/no choice or a multiple response

> **Open questions**: allow the respondent to answer freely and give greater choice, but the data that results can often be difficult to process.

General guidelines:

* **keep your questionnaire as simple as possible and short, up to four or five questions, as people are often busy**

* **select a sample of adequate size**

* **write a brief introduction which you must try to use the same way every time you begin an interview**

* **avoid too many yes/no alternative answers**

* **make sure that you have a logical sequence to your questions**

* **ask questions which will produce data that can be analysed**

* **try to get a mix of closed and open-ended questions**

* **use tick boxes for data which may be complex or sensitive**

* **as far as possible ask questions about a person's behaviour, not how they themselves think that they behave**

* **sometimes when asking for personal information, such as age, give respondents a category into which they can fit without being so precise that they become concerned. For example, a**

Figure 5.1 Asking the question

person will admit to 18 to 35 years of age but might not be willing to give an exact figure
- test out your questionnaire to see if it will produce the data that you want (piloting)
- seek approval from your teacher or tutor before you start in order to avoid insensitive questions and prevent harassment of local people by too many questionnaires.

Administering the questionnaire

This can be done in a number of ways:

1   By standing on a street and catching passers-by. If you decide to stand outside a specific service or in a shopping centre then you should ask for permission. Be prepared for a refusal. There are a number of points to consider when carrying out such a survey:

   i   be polite and explain why you want people to answer these questions
   ii   obtain a document from your school/college which explains who you are
   iii do not be upset if some people refuse to answer
   iv   look smart, be ready with a smile, and a thankyou!
   v   if at all possible NEVER work alone.

2   By going from house to house and asking questions.

3   By going from house to house and posting the questionnaire. This may be picked up later or you could leave a stamped addressed envelope.

The response rate of these methods tends to decrease in the order given, but the detail and truthfulness of responses tends to increase in the same order.

Sampling attitudes

One of the hardest pieces of information to collect during an investigation is data with regard to attitudes. There are several methods which may be used depending upon the objectives of the survey.

1   **Bi-polar tests**
   These involve establishing a rating scale based upon two extremes of attitude which are said to be 'poles apart' or with 'bi-polar' views (this is sometimes called a bi-polar semantic differential test).
   Each factor can be weighted and then multiplied by the mean score to arrive at an overall figure for one area. This is particularly helpful where a comparison of two areas is involved.

When constructing such a scoring chart it is a good idea that all the 'bad' words should not be placed on one side of the chart as this is likely to bias responses.

Projects in Practice

A good example of a bi-polar test sheet would be the following, which was used by a student to evaluate people's attitudes to a new industrial estate that was the subject of his investigation.

The layout is ATTRACTIVE   7 6 5 4 3 2 1   UGLY
Green area amounts are POOR   1 2 3 4 5 6 7   EXCELLENT
The traffic flow is CONGESTED   1 2 3 4 5 6 7   LIGHT
The place is QUIET   7 6 5 4 3 2 1   NOISY
Smells are PLEASANT   7 6 5 4 3 2 1   OFFENSIVE
Roads and ind/areas are DIRTY   1 2 3 4 5 6 7   CLEAN

2   Using a point-score scale

This is a rating scale which allows respondents to identify factors which they consider to be important. The scale usually runs from 0–4 and it is possible to add written descriptors to each number.

Projects in Practice

Sam was investigating the geography of crime within his local urban area. In order to gauge the attitudes of local residents in several different areas of the town towards crime prevention he devised a brief questionnaire which be based on the point-score scale.

Question: Which of the following factors do you think are important in creating a safe urban environment?

	Essential	Very important	Important	Not really important	Irrelevant
	(4)	(3)	(2)	(1)	(0)
Traffic calming					
Neighbourhood Watch					
Community policing					
Terraced housing					
Community centre					

3 Using a rating scale
This allows respondents to strongly agree or disagree with a statement. You can quantify the scale. As with the bi-polar test, it is a good idea to vary the questions taking both positive and negative lines in an attempt to avoid any bias in the survey.

Projects in Practice

Sally's investigation into a new shopping centre development involved looking at the attitudes of various people in and around the town in question. Part of the questionnaire that she drew up is given below.

	Strongly agree	Agree	Undecided	Disagree	Strongly disagree
The centre is attractive					
The range of shops is poor					
The centre will help the town to grow					

An example of a good questionnaire

The following material relates to a successful questionnaire used by a student carrying out an investigation into recycling in her local area. One of the aspects that Sandra wished to investigate was the success, or otherwise, of a number of recycling sites situated in various parts of the town. To get this information Sandra used the following questionnaire in carrying out face-to-face interviews with a number of people using the various sites.

In setting up the questionnaire (see page 41) and its application, Sandra explained the choice of some of the questions within the survey and her preferred method of application.

1 Question 1 was included in order to try to examine the level of success of the local authority's duty to encourage the use of recycling facilities. To make the scheme economically viable and therefore justifiable, a sufficient number of people need to use the site on a regular basis. Is each site providing a frequently used and therefore valuable service to compensate for the cost to the local taxpayer?

2 Question 3 was put in to identify the catchment of each recycling point. If sites are too widely dispersed this might deter people from using the service.

3 Question 5 addresses one of the issues involved with recycling.

4 Question 7 looks at the willingness, or otherwise, of local people to participate in alternative schemes. If the local authority had ideas of trying new schemes, they would have to be seen as more convenient by the population.

5 Sandra attempted to set up a stratified sampling technique by including in her sample as wide a cross-section of the community as was possible. When she had collected 50 per cent of her total at each site she reviewed the results in order to try to target those that she felt were under-represented.

6 At each recycling point, the target sample was 50 respondents. In some cases it meant that she had to return on another day in order to achieve her sample.

Sandra also planned to carry out another questionnaire asking people about their attitudes to recycling and other forms of waste disposal. This she did in the town centre using a stratified sampling technique.

5.9 Point sampling

You would use sampling of this kind when the whole area or number of features to be looked at is large and the only feasible way of carrying out the investigation is to obtain results from a few points. Examples of such surveys include:

- land-use surveys, particularly in rural areas
- selection of pebbles on a beach when carrying out investigations into longshore drift
- vegetation sampling on a slope
- selection of sites for soil investigations.

Point sampling is carried out by use of some sort of grid. The grid, for example, that is to be found on Ordnance Survey maps is ideal for this purpose or you could construct your own, but in the field it is useful if a quadrat is used. A quadrat is a frame enclosing an area of known size which in most cases is one square metre. The area of the quadrat is subdivided by use of wire or string and, with a numbering system, sampling of such features as pebbles or plants can be carried out using random or systematic methods.

QUESTIONNAIRE AT RECYCLING POINT

Location ....

Excuse me, I am a college student carrying out an investigation into recycling in this borough as part of my A Level geography course. Could you help me by answering a few questions please?

1 How often do you use this facility?
  a More than once a week ❑
  b Once a week ❑
  c Once a fortnight ❑
  d Other (please state) ❑ ....

2 Do you use this recycling point because
  a It is close to home and an easy way to dispose of rubbish? ❑
  b It is easy to use when going to shop/work? ❑
  c Or do you have to make a special journey to use this facility? ❑
  d Other (please state) ❑ ....

3 How far do you travel to this site?
  a under 1/2 mile ❑
  b 1/2 to 1 mile ❑
  c 1–2 miles ❑
  d over 2 miles ❑

4 How did you travel to the site?
  Walk ❑
  Car ❑
  Cycle ❑
  Other ❑

5 Do you find any of the following problems at this site?
  a Noise and fumes from traffic coming on to the site ❑
  b Noise and fumes from lorries taking away the materials ❑
  c Litter ❑
  d Vandalism ❑
  e Smell ❑
  f pests (e.g. flies) ❑

6 Do you think that the value of recycling outweighs any environmental problems that you have identified in the previous question? Yes ❑ No ❑

7 Would you prefer a separate bin with compartments at your home rather than bringing waste to this collection point? Yes ❑ No ❑

Sex .... Age: .... under 21 ❑ 21–30 ❑ 31–60 ❑ over 60 ❑

Thank you very much for your help.

A B C D E F G H I J K L M N O P Q R S T U V W X Y Z

Projects in Practice

Raad was undertaking a survey to ascertain the direction of longshore drift on a coastline. To do this he needed to look at pebble size along the stretch in question and with the numbers involved it was obvious that he would have to carry out some form of sampling. He made several decisions at the start of the survey:

1   That he would only sample one type of rock which he could recognise as other types may be moved at different speeds and therefore he would get some consistency at each site. For his survey, Raad chose to sample granite pebbles as this was one rock he could recognise and it was a common rock in the area.

2   He would use a quadrat sampling technique but would choose the sampling sites systematically. As the beach was between one and two kilometres in length, he chose to use the quadrat every 100 metres. This would give him around 15 sites.

3   Sampling would be carried out at the same distance up the beach at each point, as sizes of pebbles can vary in a transverse direction as well as longitudinally.

4   When using the quadrat at each site, Raad had a choice of using either a systematic or random sample. The operation of each system is shown in Figure 5.2 below. He eventually chose to select the pebbles using a systematic system based upon 16 points within each quadrat as this would give a large enough sample bearing in mind that all the pebbles at the selected points might not be granite.

The random points were selected by Raad using the following random number table:

21	18	43	28	24	18	60	67	39	62
75	50	05	49	03	06	11	34	84	71
95	71	89	31	38	67	23	42	26	65
78	15	22	15	69	84	32	52	32	54
93	29	12	12	27	30	07	55	91	87

Raad used the numbers as follows:

Continued

21 and 18 were used for the first plot. 21 became 2.1 easting and 18 became 1.8 northing.

For the second, 43 became 4.3 easting and 28 was changed to 2.8 northing; 24 and 18 were used for the third plot, 24 became 2.4 easting and 18 translated into 1.8 northing and so on through the table until 16 plots had been made.

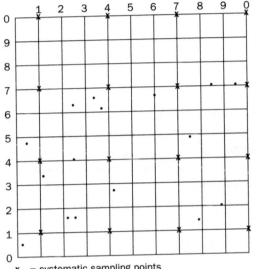

x = systematic sampling points
. = random sampling points

Figure 5.2 The operation of systematic and random sampling

5.10 Surveys

So far the information that we have shown you how to collect has been done mainly through asking questions of other people. Material for your investigation though, can be collected by you carrying out your own research in some form of survey. There are a large number of different types of survey that you could use and below we have listed some of them with details, in a few cases, of how to collect the information.

Land-use surveys

These can be done for both rural and urban areas and must always be carried out with a clear purpose in mind. The best way to carry out an urban survey is to group land-use into classes where groups of similar land-use types can be identified. One such grouping is given below:

Grouping	Division of one group
Residential = R	RT – terraced house
Industrial	RD – detached house
Commercial	RB – bungalow
Entertainment	RS – semi-detached house
Public buildings	RF – flats
Transport	
Services	
Open space	
Vacant buildings can also be recognised	

It is also possible to divide up each of the above groups:

- **This will depend, however, upon the detail that you wish to include in your survey which in itself will depend upon your overall objective.**

- **A division of the residential group could run as shown above right.**

Land-use transects

If the urban area is too large to investigate in the time available then a shorter form of survey would be to look at a sample of the land-use of the area in question by taking an urban transect. A transect is a technique which essentially involves taking a slice through the urban area to see how the land-use changes from one part of the area to another. Most transects start somewhere near the centre of the town or city and run in a radial manner along roads to the urban fringe if it is required that you go that far in your survey. They are often used for comparison to show how one side of an urban area differs from another or perhaps, in another context, to trace the extent of the central business district (CBD) by plotting where such central function cease. Other surveys that could be taken along a transect include upper floor use, building height, building condition, a number of environmental investigations, including noise and pollution surveys, and some climatic/weather surveys such as temperatures across an urban area or a valley.

Environmental surveys

These can be carried out as some form of appraisal or assessment and can generate important information about the environment under investigation. The use of a point score scale is recommended where both positive and negative observations can be made.

Projects in Practice

Sarah was involved in an investigation where she was looking at various beaches in an area of south-west England, all of which had some tourist impact upon them. To make a detailed comparison of this kind she devised the following assessment sheet which was completed for each locality

Title: Beach quality survey

Location .... GR .... Date ....

Weather .... State of tide ....

Score							Criteria	Comments
−3	−2	−1	0	+1	+2	+3		
							Water quality	
							Beach composition	
							Beach gradient	
							Width of breach	
							Type/height of waves	
							Access/ car parking	
							Litter/oil/tar	
							Sea views	
							Inland views/ landscape	
							Dog numbers/ dog mess	
							Quality of facilities	

Continued

Score							Criteria	Comments
-3	-2	-1	0	+1	+2	+3		
							Visitor density	
							Variety of habitats	
							Wildlife in beach area	

Assessment

You could carry out a similar survey when looking at the environment quality of an urban area. For such a survey you will again require to set up an index scheme and the example below does not scale the assessment into positive and negative, as the beach survey above but, similarly to the bi-polar test already shown, uses a point index where 1 represents the worst possible case and 10 the best possible.

Other forms of pollution surveys could include investigations into the level of noise, air and water pollution. It is possible to obtain instruments and chemical testing kits in order to carry out such surveys but you can carry out simple tests using your own observations or at the very most some simple equipment.

Projects in Practice

Simon's environmental quality survey was carried out in a section of the inner city of the area in which he lived. He chose a number of categories in his survey chart, all of which he could observe as he investigated the area. He also considered including a category of crime but realised that he would have to try to get such information from the police or the local newspaper/library archives; he doubted that he would be able to witness it in practice! Another way of carrying out such a survey is to use a bi-polar test based upon interviews with residents in order to see how they perceive their own environment.

Continued

Title: Environmental quality index

	a lot of 1	2	3	4	5	6	7	8	9	10	none
Run down housing											
Derelict land											
Heavy industry											
Traffic											
Litter/ vandalism/etc.											
Mechanical noise											
Air pollution											
Untidy play areas											
Untidy gardens											

Simon then totalled the scores and found a percentage (in this case of a maximum 90) that gave an index based upon 100 and enabled him to make comparisons between areas within the inner city.

Noise pollution surveys

You could make your own observations but it is difficult to measure noise accurately with the human ear. A simple solution is to make sure that one person carries out all of the tests and that you devise a scale, such as the one below, with which you can compare areas.

Level	Decibels	Observations	
1	Under 10	Still	No noise other than the sound of breathing
2	10–20	Quiet	Branches and leaves moving in the wind. No human noise
3	21–40	Normal	Distant noise such as traffic. No immediate source of noise
4	41–70	Loud	Disturbing/distracting noise such as heavy traffic
5	71–100	Very loud	Unpleasant level of noise possibly causing vibrations
6	Over 100	Severe	Hurts ears; need to move away from source of noise

Air pollution surveys

There are a number of simple tests that you can carry out but for the one shown below you only require a roll of clear sticky tape which you use to take samples from the same type of surface in a number of different localities. For the survey shown below the tape was used on the surface of street furniture in a CBD.

Level	Observations
1	No obvious signs of pollution on tape
2	A few traces of pollution shown
3	Visible coating of pollution on sticky tape
4	Significant evidence of pollution traces
5	Evidence of high levels of pollution. The tape is heavily coated

Projects in Practice

Mark was looking at his local river as it flowed through the centre of the urban area on which he was carrying out a pollution investigation. He decided in the first place to survey the river in a number of sites and devised the following index system which he could apply at each of them.

Water pollution visual survey index

polluted	1	2	3	4	5	clean
very murky water						clear water
stones covered with scum						clean stones
much grey sewage algae						no sewage algae
much scum, froth, oil						no scum, froth, oil
lots of rubbish						no rubbish
some strange colours						no strange colours
choked with weed						very little weed

Site . . . .   Total points . . . .

Water pollution survey

Many tests involve smell, colour, amounts of foam, levels of solids, and oxygen levels. By just looking at a stretch of water you will gain a reasonable idea of how healthy it is if you know the signs that you should be looking for. As with previous surveys shown, create an index for your assessment.

Other types of surveys that could be carried out in an urban context include:

1 Building condition surveys

This survey could be part of one of the above environmental surveys or it could stand on its own right. This again could be used with an index, perhaps you could call it an Index of Decay.

2 Shopping quality surveys

Here you could use an index to find out the type and quality of shops in an area or you could turn it into an investigation into the environmental qualities of shopping areas. You could take on a similar type of survey for offices or other services.

3 Convenience surveys

This could involve looking at the service availability in terms of distance in a residential area such as shops, schools, doctors/dentists, parks, leisure centres.

4 House price survey/Land value surveys

You could make a survey of the prices of houses for sale in an area as this should show relative land values. You could do the same thing by finding out the value of property for local tax assessment but this is somewhat more difficult a task than taking house prices from newspapers or estate agents. In a CBD you could look at a comparison of land values through the area. To do this you will need access, as indicated, to the local authority's valuation office where details of the value of land for local taxation purposes are kept and open to inspection. The value of each property will have to be converted into a unit per square metre of ground floor space to overcome distortions caused by the different size of buildings. From such information the Peak Land Value Point (PLVP or PVI – Peak Value Intersection) can be determined.

5 Pedestrian flow surveys

People walk inside the CBD of a town or city and by measuring this flow it is often possible to find the centre of the area and, perhaps more importantly, its limits. Such a survey should be carried out with a number of your friends and colleagues and there are a number of points to remember when carrying out the exercise:

i   try to carry out the exercise in mornings and afternoons in order to avoid the movement of office and shop workers

ii   do more than one survey, either to get an average or to contrast different times of the week. Remember that professionals will carry out surveys on each day of the week in order to establish the general pattern of movement

iii   in the areas that you consider busy, you will need two counters

iv   stand at one side of the street or pedestrian way. If there are two of you, the best position is standing back-to-back in the middle so that you get an accurate picture by both of you not counting the same person (this may, of course, not be possible on actual streets)

v   shopping centres are usually private property so it is better to seek permission if you intend to conduct a count at one of the doorways.

6   Traffic flow surveys

This can be carried out by measuring the traffic flowing past several survey points or you could weight the survey by giving each type of vehicle a different score (i.e. car 2 points, vans, etc. 4 points, heavy vehicles such as lorries and buses 6 points).

7   Shop location surveys

Within a CBD there is often a distinct locational pattern of land use. Certain types of shops (and services, offices) tend to cluster while others are usually more widely dispersed. Firstly you should carry out a normal land use survey and then select a category which you wish to examine, shoe shops for example. Measure, the distance from each shop/office to the one which is closest to it in that category it and then carry out a Nearest Neighbour Analysis which is covered in the analysis section of this book. You could also calculate an index of dispersion.

5.10 Collecting data for surveys in physical geography

River surveys

There are a number of measurements that can be made within the valley and channel of a river and that can be used in the calculation of several fluvial features such as discharge, load, friction and efficiency.

1   How to measure the speed of a river

The best and easiest method is to use some sort of flow vane which will enable you to obtain either a direct reading or be able to calculate the speed on a conversion chart. Without a flow vane it is possible to calculate the speed by using a piece of wood or a cork as a float. The procedure is as follows:

i   measure a ten metre section of the river

ii   measure the time it takes the float to cover the distance

iii do this several times and at various points across the channel

iv   find the average time – this is the surface speed

v   multiply by 0.8 to find the true speed of the river which will be below the surface and about the point that you used for your calculations with the flow vane.

2   Measuring the cross-section area and wetted perimeter

Firstly run a tape across the river along the water line from bank to bank. Along this tape calculate the depth of the stream at every 50cm. Once calculated, transfer this information onto a piece of graph paper making sure that you use the same scale on both the vertical and horizontal axes. Having produced the cross-section diagram you simply count the squares (whose size you already know) and a simple calculation of multiplying number with size will give you the cross-sectional area. You can also measure the wetted perimeter from such a diagram.

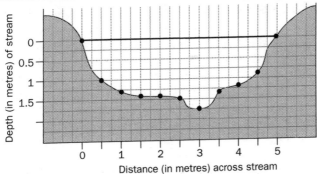

Figure 5.3 River cross-section and wetted perimeter

3   How to measure the discharge of a river

Once you have found the speed (in metres per second) and cross-sectional area (in square metres) you simply multiply them to give the discharge in cubic metres per second or cumecs.

4   Calculation of river gradient

You use a clinometer and siting pole and carry out the same technique as in slope measurement (see section below on slopes).

5   Measuring variations in bed load shape and size

At each measuring site select ten pebbles, which can either be done by random selection or by use of a quadrat. Measure the long axis of each pebble to give you the size. The angularity of the stones can be assessed by using the Shape Chart (Figure 5.4).

Another feature that can be calculated is the Index of Roundness. To do this you will need to take the following steps:

i   measure the long axis and then divide by two to give the horizontal radius

ii   measure across the pebble at right angles to the long axis and divide this figure by two to give the vertical axis

iii calculate the index of roundness as follows

$$\text{Index } (R) = \frac{\text{horizontal radius} + \text{vertical radius} \times 100\%}{\text{long axis}}$$

a pebble with $R = 100$ would be perfectly round.

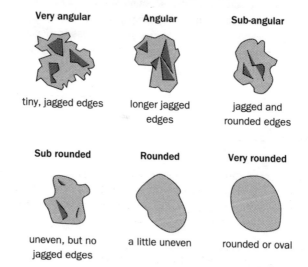

Very angular	Angular	Sub-angular
tiny, jagged edges	longer jagged edges	jagged and rounded edges

Sub rounded	Rounded	Very rounded
uneven, but no jagged edges	a little uneven	rounded or oval

Figure 5.4 Shape chart

6 **Measuring sinuosity**

Measurements should be taken as follows:

i mark a position half way round the two sides of a meander – between A and B and between B and C. Call these points X and Y – see Figure 5.5.

ii measure the straight distance across from X to Y

iii measure the distance around the bend from X to Y

iv calculate $\dfrac{\text{Distance along the river X–Y}}{\text{Distance in a straight line X–Y}}$

This is the index of sinuosity – the nearer to 1, the straighter the river.

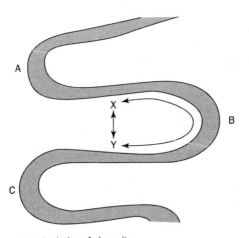

Figure 5.5 Measuring the index of sinuosity

Surveys on slopes

1 **Measuring steepness**

To obtain the profile of a slope you will require ranging poles, a clinometer and a measuring tape. You should measure each section up to a break in the slope. If you cannot obtain ranging poles, then use a friend as shown in Figure 5.6 on page 54.

2 **Vegetation surveys**

This type of survey requires the use of a quadrat which is placed on the slope at a number of selected sites. Firstly decide the method of choosing your sample points, probably systematically, and then place the quadrat on the ground; to ensure it is done randomly the best way is to close your eyes and then cast it away from you and onto the ground. Within the quadrat you may choose a random or systematic method for identifying sample points. Surveys that can be made include:

i   vegetation type and thus diversity of plant species

ii height of vegetation

iii percentage of vegetation cover (it is possible here to use the quadrat squares which can be counted for each vegetation type or vegetation cover compared with bare ground).

Information of this kind can be displayed on a kite diagram.

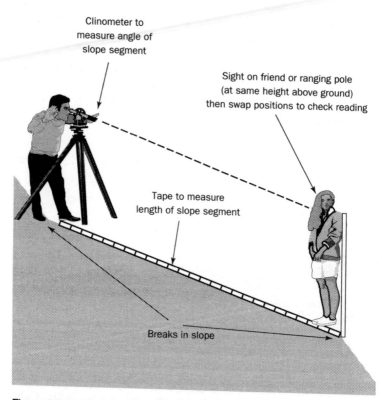

Clinometer to measure angle of slope segment

Sight on friend or ranging pole (at same height above ground) then swap positions to check reading

Tape to measure length of slope segment

Breaks in slope

Figure 5.6 Using a clinometer

3 Soil surveys

Sites for survey can be chosen exactly the same way as for vegetation. At each site a soil pit can be dug or the soil tested by means of an auger which is in effect a very large hand drill. Soil profiles can be drawn showing the horizons within a soil pit and various tests can be carried out including:

i soil acidity – use a soil testing kit in the field. A barium sulphate testing kit is probably the best to use here but you can purchase meters or a simple pH set from a garden centre. At such centres you could also buy a soil chemical testing kit which should give you readings for nitrates, phosphates and potash

ii soil texture – a rough guide in the field would be to feel the texture with your fingers: sand feels gritty; silt feels much smoother, often silky; clay tends to stick to the fingers. You can also test a sample by seeing how long particles take to settle in water or you could use sediment sieves where your sample is held by an increasingly fine mesh as it moves down the containers

iii organic content – take a sample and then dry it overnight in an oven. Crush the soil and weigh it and then place over a bunsen burner which will burn off the organic matter. Weigh the sample again to find:

$$\frac{\text{weight loss after burning} \times 100\%}{\text{weight of dry soil}}$$

iv soil water content – weigh your sample and then dry it overnight, then reweigh. Water content will be:

$$\frac{\text{weight loss on drying} \times 100\%}{\text{weight of wet soil}}$$

Coastal surveys

1 Measuring longshore drift

One method of observing longshore drift is to paint a number of pebbles in a distinctive colour and then place them on the beach. When returning at set intervals you should be able to chart the progress of your indicators along the beach. You could also measure the size or roundless of pebbles along the beach to see if there are any significant changes which would indicate the direction of longshore drift (for details of an example of such a survey see Projects in Practice on page 42). If groynes are present on the stretch of coastline then you could observe their location, the deposition around them and the wave height either side.

2   Sand dune surveys

These represent a dynamic environment for certain plants and animals and a number of relationships can be established. Sand dunes are also areas where the influence of man can be seen on the physical environment.

Projects in Practice

Judith wished to undertake an investigation into the sand dunes of her local coastline. The two main areas of this work would be, firstly to see how the dunes changed with distance inland, and secondly to look at the impact that man has had upon the area. To investigate the changes in the dunes she selected a line of survey through the area (a dune transect) and then set up a number of hypotheses:

- the community of plant species will become more diverse away from the sea, culminating in some form of climax community
- the organic content of the dunes will increase inland
- freshwater content will increase inland
- dune size will vary from the small foredunes inland.

Judith intended to carry out this first part of her investigation by measuring slope size and angle through the dunes (ranging poles, clinometer, tape measure), vegetation sampling (quadrat); soil testing (barium sulphate kit, auger), and a small microclimate survey (hygrometer, anemometer).

Microclimate surveys

Small scale observations can be made in your local area either as an investigation in its own right or as part of a larger survey, such as the one shown in the above Projects in Practice. Surveys can be both areal and/or temporal, such contrasts being made between day and night, from day to day or season to season. Equipment will include: a rain gauge, maximum and minimum thermometers, wet and dry bulb thermometers (hygrometer), and an anemometer.

Glaciation surveys

The study of glacial deposits can be carried out using a technique called till fabric analysis. This is based upon the idea that stones within the ice will become orientated in a direction that presents minimal resistance. This means that they should be found with their long axis parallel to the direction of ice movement, therefore if you measure the long axis orientation of a number of pebbles in a glacial deposit you should be able to establish the movement of ice that laid down the deposit and possibly the source area of that ice. Once you have established the direction it should also be possible to determine the glacial history of the area of study. The procedure is as follows:

1   **Select a number of stones from within a glacial deposit and, using a compass, note the orientation of the long axis.**

2   **Group the stones into classes of ten degrees and then plot this information onto a rose diagram. Each stone will have two orientations to plot, which of course will be opposites i.e. 10 degrees and 190 degrees also 90 degrees with 270 degrees.**

3   **Once you have entered all the classes, you join the points to form a typical rose diagram which can be shaded. The classes which reach out furthest from the centre indicate the most likely direction of ice movement. Relate your sampling point to a map of the area and observe how your findings fit into the landscape.**

4   **Stones within glacial deposits also show a tendency to dip up-glacier i.e. towards the source of ice. As well as measuring the orientation, also record the direction of dip and this information can be placed on a rose diagram. This should confirm your original finding of the orientation survey.**

5.11 Using a recording sheet

When carrying out physical surveys, a good recording sheet which can be duplicated for use at each site is essential. The following pages include recording sheets for three types of investigation and whilst all investigations have slight differences, such sheets could be of great value to you in the field after a little adaptation to your particular circumstances.

River Survey Sheet

Name of river ....   Grid reference ....

Date ....

Weather ....   Previous weather ....

Width of river ....   Wetted perimeter ....

Cross-section measurements

Distance

Depth

Height of banks above water   1 ....   2 ....

Time taken for float to travel 10 m downstream

1 ...... seconds   2 ...... seconds   3 ...... seconds   Average ......

If using a flow vane measure speed at 10 cms below surface
...... metres per second

Composition of river bed (stony, sandy, etc.)

Collect and measure the long axis of ten randomly chosen pebbles

1 ....   2 ....   3 ....   4 ....   5 ....

6 ....   7 ....   8 ....   9 ....   10 ....

Average ....

Evidence of erosion or deposition ....

Land use on banks ....

Evidence of previous high river levels ....

Coastal Survey Sheet

Name of area .... Date of survey ....

Sea

Compass direction of incoming waves ....

Number of incoming waves per minute ....

Land*

Description of coast (beach, cliffs, etc.). Include a field sketch below

....

Beach composition (rock, shingle, etc.). Estimate cover if more than one type

....

Evidence of coastal erosion ....

Evidence of coastal deposition ....

Measurement of pebble size (along longest axis)

Sample number	1	2	3	4	5	6	7	8	9	10
Near sea										
Back of beach										

* You could add a human impact assessment such as that given in the Projects in Practice on page 45.

A B C D E F G H I J K L M N O P Q R S T U V W X Y Z

Soil Survey Sheet

Location ....   Grid reference ....

Recent weather ....   Date of survey ....

Sample of soil from 5 cm under surface

Colour ....

Texture(sandy, clay, etc.) ....

How wet is the soil? ....

Soil acidity (pH testing) ....

For the site

Soil temperature ....

Soil depth (use an auger) ....

Organic content – is there a clear humus layer? ....

What animal life is present? ....

What is growing in the soil? ....

Land use ....

If possible draw the soil horizons from the material brought up by the auger – if present. You could also dig a soil pit if you have the strength and, most importantly, obtained permission.

5.12 Collecting information through interview

You have already seen how it is possible to collect information by using a questionnaire. With certain projects, some of the information that you require may be obtainable from newspapers where people are expressing opinions on the subject of your investigation to a journalist (see Secondary Collection, page 65). It may be necessary, though, to conduct interviews in order to see how people stand on an issue or how they act in certain circumstances. Interviews with farmers, for example, may help to account for some of the changes in the land-use pattern of a local area. Where there is a conflict that comes within the scope of your investigation then it is essential that you obtain the views of the parties most affected. In such cases you may have to go beyond the formal nature of the questionnaire as some of the questions that you may wish to ask could be different for each respondent and group. If you cannot meet people face to face then you may have to write to the person or group for their opinions.

Interviews give you an opportunity to ask more detailed questions and to follow up points of interest. They are somewhat time consuming, so are best used sparingly. Send a well presented letter which will give a good impression: do this early in period of your investigation as it may take some time to set up a meeting. Always prepare well for an interview, people do not like wasting their time. Before you go, prepare a list of the crucial questions that you need to ask and make sure that these are direct questions which only the interviewee can answer and certainly not questions where you could have discovered the information for yourself. Do not overstay your welcome and try to keep a record of what was said, either by writing notes during the interview or immediately afterwards. If possible you could record the interview but always ask permission before starting. Quotations from interviews can add a lot to your final presentation but always check accuracy and ask permission from the interviewee.

Projects in Practice

Janet's investigation concerned the levels of congestion from vehicular traffic that were being caused in her home town which was a holiday resort on the south coast of England. One of the proposals was to construct a Light Rapid Transport (LRT) system, and in particular put some of it in a tunnel under the harbour. Janet decided that as one source of data she should interview some of the parties involved to see where conflict existed. These would include:

- the chairman of the County Council LRT project committee
- the managing director of the local bus company
- the manager of the ferry company operating in the harbour
- residents along the routes affected by the proposals
- an environment group, possibly Friends of the Earth
- tourists coming to the town.

Continued

From her interviews she was able to construct a conflict matrix through which she would be able to see the extent of the problem. The local bus company, for example, could see that some of their business would reduce, but if they changed their current network to fit in with the LRT then many passengers would perhaps prefer to travel with them as fares would undoubtedly be cheaper on the buses. Janet therefore completed the following conflict matrix:

	Bus company	Ferry company	Friends of the Earth	County council	Residents	Visitors
Bus co.	✓	✔	○			
Ferry co.	✓		✗			
F. of the E.	✔			✔		
C. c	○	✗	✔		✓	✔
Residents				✓		
Visitors			✔			

Key

✗ minor conflict   ✓ weak agreement   ○ some conflict/agreement

✗ major conflict   ✔ strong agreement   no relationship

5.13 Using field sketches and photographs

Field sketches and photographs are both excellent ways of recording in your investigation exactly what you have seen. Field sketches enable you to pick out from the landscape the features that you wish to identify and perhaps comment upon. This is particularly the case when dealing with an investigation in physical geography. Coastal geomorphology often lends itself best to this method, but there is no reason why you cannot make and present sketches from within an urban area. You should not worry though, if you are not particularly artistic as it is far more important in geographical investigations to produce a clear drawing with good and useful labels.

Not every student is confident enough to produce field sketches, therefore if you feel the need to show detail from the area under investigation then take photographs. Do not take photographs for the purpose of providing attractive space fillers within your write-up in the mistaken belief that they will make your final presentation 'look good'. You need to select very carefully the images you want to show and remember to take the picture so that it shows the necessary detail to the reader. If possible you should take a mixture of shots showing both the wider area and the closer view with details. The most important point about sketches and photographs is that when you come to the stage of presentation they must be annotated or labelled. You may have already done this in the field with your sketch but a photograph is only really valid if you point out to the reader the major features which you have observed, these being points that are important in your investigation. You should include the photographs at the relevant point in your presentation. Do not include them all together as a large block as this makes it more difficult for the reader to see the reference point and thus the validity of the method. Remember that field sketches and photographs can provide evidence like any other fieldwork method.

Projects in Practice

Ben was conducting traffic surveys as part of his investigation into the need for some form of relief road to take traffic out of the centre of the town where he lived. The present traffic densities varied tremendously between different parts of the town and at different times of the day. As well as conducting traffic counts at various times, Ben decided to take a number of photographs to give a visual record of the flows. He was also able to show the type of traffic using the roads at various times throughout the day and where certain vehicles, such as lorries, were concentrated.

SKETCH OF THE BEACH NOURISHMENT IN PROGRESS AT GRID REF. 195348

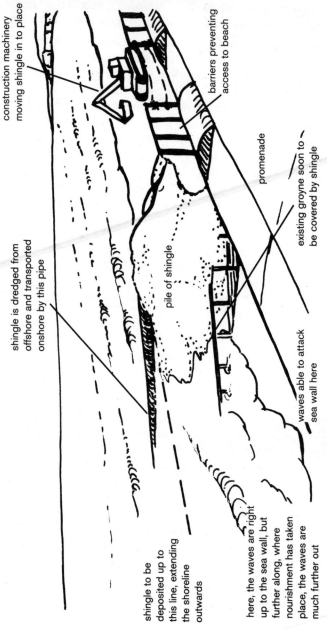

construction machinery moving shingle in to place

barriers preventing access to beach

shingle is dredged from offshore and transported onshore by this pipe

promenade

existing groyne soon to be covered by shingle

pile of shingle

shingle to be deposited up to this line, extending the shoreline outwards

here, the waves are right up to the sea wall, but further along, where nourishment has taken place, the waves are much further out

waves able to attack sea wall here

Figure 5.7 Example of a field sketch made on a coastal survey

6 Collecting the data: secondary sources

Secondary research means gathering data which has already been compiled in written, statistical or mapped forms. The research inevitably will not always have been tailored to meet the specific requirements of your own investigation, but for the geographer there are a wide range of sources of secondary material that can be accessed and this chapter explains about the data available and how it may be used. On page 201 of this handbook you will find details of addresses from which you could obtain certain material.

6.1 The importance of secondary data

Data collection in the field, by its very nature, can only supply a limited amount of data for some investigations. It is therefore important that the acquisition of secondary data is carried out, particularly where your investigation has a temporal context. If, for example, you are carrying out investigations into pollution or traffic, the local authority may have information on previous surveys which could be accessed. Secondary material can be very useful in providing a context in the initial stages of an investigation and it can also play an important part in the discussion and explanation of the data collected in the field. In the area of human investigations many students combine their field data with material that they obtain from map, census, newspaper, local authority and other secondary sources. This will give you a much wider database for analysis and sometimes makes it possible for the results from first hand collection in the field to be compared with other results obtained from an analysis of secondary data and senior examiners certainly welcome this type of approach to your investigation. Good examples of where secondary data could prove a valuable aid in explanation include the following:

- **planning data relative to studies in residential areas and in the CBD**
- **secondary data on the drainage basin and information with regard to the rainfall regime in river basin studies**

- **detailed geology maps and information when studying coastlines**
- **synoptic chart data in micro-climate investigations.**

It is important that secondary data should usually be questioned in respect of the quality and accuracy of the material. It is also good practice for you to be aware of how to best use such information.

ACCURACY AND REFERENCING

When using secondary material it is important to consider the accuracy of the source and if information could be biased. Remember to keep details of author, title, publication, etc., as these will have to be incorporated within your final work. You will certainly want this information for your bibliography and it is a good technique to make reference to secondary sources in your text by the use of footnotes or brackets.

CHECKING THE FORM OF DATA

There are problems that could occur if you are relying on extensive data from bodies such as health or police authorities. Information or statistics that are available frequently fail to meet needs, as the bodies do not collect data at the same scale as you require. Crime figures, for example, are not usually available for individual streets. It is important at an early stage in your planning that you check not just the availability of data but also its form. You do not want to find at some distance into your investigation that the secondary data does not match up with your primary collection.

6.2 Sources of material

CENTRAL GOVERNMENT DATA

Government data covers a wide range from economic data through to population material and crime statistics. You should find in most libraries volumes of published data such as *Annual Abstract of Statistics, Population Trends, Social Trends, Economic Trends and Monthly Digest of Statistics*. All of these are published by the Office for National Statistics and are available through Her Majesty's Stationery Office (HMSO). It is better to use the local library, if possible, as some of these publications can be expensive to buy.

- *Annual Abstract of Statistics* **will give you figures over a wide range of topics enabling you to make comparisons with previous years – a good example would be car ownership over the last decade**

- *Population Trends* presents statistics on population, population change, births and deaths, life expectancy, migration, marriages, divorce and related areas

- *Social Trends* contains data about people's lives and habits such as how many people own a car, go to the cinema or play football

- *Economic Trends* contains all the main economic indicators such as GDP and employment/unemployment

- *Monthly Digest of Statistics* is the best source of up-to-date information on population, production, employment and the weather.

Another option is to write to the press offices of government departments who can provide useful material.

Government agencies can also be contacted such as the Highways Agency, who are responsible for all new road developments and improvements.

LOCAL DATA

There are a number of sources that could be accessed locally including the following:

- the local authority who will have information on planning and on population levels in local areas – local government statistics and publications are usually available in your local library but sometimes you need to contact the relevant department of the local authority

- the electoral register which could be useful for identifying 17/18-year-olds for a survey

- local estate agents which will give you information on property prices

- *Yellow Pages/Thompson Local* which lists local firms categorised into business types

- local libraries, which apart from containing the national information given above, will also have statistics for local population taken from the national census. This information is usually broken down into areas as small as electoral wards and may contain details on only a few thousand people. This information is known as Small Area Statistics (SAS) and contains material on age, ethnic and employment structures of your local district. Local libraries are also useful for back numbers of local newspapers and other historical information such as details of a previous census (even going back into the nineteenth century)

- the local Chamber of Commerce or Training and Enterprise Council (TEC) could put you into contact with local firms – the Chamber of Commerce may even publish a local directory

- the Health Authority for your area will publish statistics, probably broken down into electoral wards, of births, mortality rates and information relating to living conditions, such as persons per room in households

- local newspapers will have information on issues such as planning, traffic, crime and health. This also will be a good source of opinions or for getting information on people/groups to interview. The newspaper offices will have back numbers but the best source is the local library where the information is usually stored on microfilm unless it has been updated to CD–ROM

- photographs from local newspapers may help you in looking at land-use and other changes. Many local libraries have very good photographic archives which you can access and photocopying opportunities may also be available on the premises

- local action groups that are concerned about the impact of particular developments and may provide campaign literature

- directories such as *Kelly's* and *Marshall's*, which contain data going back to the nineteenth century, will give you details of the type and location of businesses.

NATIONAL SOURCES (OTHER THAN GOVERNMENT)

Apart from government material, there are sources that can be accessed which have a greater range of information than that found at the local level. Such information though, may very well be about the area in your investigation. Some of these sources are:

- national media, including newspapers, magazines and television and radio programmes which may cover the local situation but might also convey information on national policy (on road building, for example)

- company publications could be useful in such investigations as those concerned with shopping centre, superstore and housing developments. Many large companies employ an education officer or adviser and it is to this person that initial enquiries should be made. The press officer in such a firm is another point of contact. Do remember that such people are busy and a reply to an enquiry may not always be immediately possible

- information from charities, action groups and national organisations such the Wildlife Trust, Shelter, English Nature, English Heritage and the Countryside Commission
- National Park and Country Park publications
- environmental pressure groups such as Friends of the Earth, Greenpeace and the Council for the Protection of Rural England
- the Meteorological Office.

GEOGRAPHICAL SOURCES

These are the traditional sources that you will have probably have been using in your school career. They include:

- maps and charts –there are many different types of map that you could use from local town centre maps to geological, land use and shop type. Sources include the Ordnance Survey, the Geological Survey, local authorities and Charles Goad, whose maps show the detailed ownership of town centre commercial buildings
- geographical magazines and journals – the best of these are: *Geography* and *Teaching Geography* (published by the Geographical Association); *Geographical Magazine* (published by Campion Interactive Publishing); *Geofile* (published by Nelson Thornes); *Geography Review* (published by Philip Allan Publishers).

THE INTERNET

If you have access to the Internet there could be some useful starting points but use of the Internet is not really necessary in geographical investigations. There may be many sites that could be useful to you but do not spend hours clicking in the hope of finding something. A visit to your local library would be far more effective.

On the Internet you can find information from government sources such as the Central Office of Information and the Office for National Statistics (ONS).

Details on your nearest Chamber of Commerce are also available along with material from the *Thompson Directory* which will give you information on the location of local businesses.

✓ Checklist 2

Is my investigation adequate?

The data has now been collected and you are now within the process of analysis. These are the questions that you should be asking of your investigation.

	Yes	Partly	No
1 Has my preliminary analysis suggested that I have sufficient data to take this investigation through to a reliable and worthwhile conclusion?	☐	☐	☐
2 If more data is needed can it be collected in the same way and be comparable to what I already have?	☐	☐	☐
3 If I am using a set of techniques in analysis, will each one advance my understanding and interpretation of the data?	☐	☐	☐
4 Have I analysed the data as fully as necessary to achieve my aim and have I used the appropriate techniques?	☐	☐	☐
5 Have I constructed the diagrams and other illustrative material that I plan to use when I write up the investigation	☐	☐	☐

Presentation of results

It is important that the results which you want to include in your study are relevant and are presented using appropriate techniques. There are a number of graphical, cartographic (map based) or tabular methods which could be used – the methods selected should be appropriate to the purpose of the study. Presentation methods are rarely used as an end in themselves, they are an important element in the analysis of results (see Chapter 8) and therefore they should be selected and applied to the data to enable you to describe changes, establish differences or identify relationships.

It is tempting to use as many different presentation techniques as possible; this can lead to similar data being presented in two or three different ways for no purpose other than to show that you know how to construct several forms of graph or diagram. This can be counter-productive in that some of the techniques may not be entirely suitable for the analysis needed. Mark schemes award credit to candidates who use a *suitable* range of techniques which provide potential for analysis. It is much more effective to use techniques which are appropriate to the specific task in relation to the type of analysis which you are undertaking.

7.1 Selecting suitable techniques

The choice of technique depends upon what aspect of the study you are trying to display; generally this will be either:

- **description or identification of differences e.g. comparing one area with another, or identifying changes/trends over time**

- **describing spatial patterns e.g. variations in levels of income in an area**

- **sorting data according to characteristics, classification e.g. soil type according to percentage sand/silt/clay**

- **identification of relationships between two sets of data e.g. environmental quality and distance from the city centre.**

Table 7.1 Methods of presenting results

	Intended use	Graphical	Cartographic
1	Identifying differences	Line graphs (arithmetic and logarithmic) Frequency curves Lorenz curves Polar graphs Pie graphs and bar graphs Proportional symbols Long/cross sections Dispersion graphs/box and whisker plots Histograms	Graphical methods such as pie/bar graphs or proportional symbols can be plotted on a base map to show spatial variations in the differences which are being studied
2	Describing spatial patterns		Isopleth maps Choropleth maps Flow diagrams/desire lines
3	Classification	Triangular graphs	
4	Investigating relationships	Scattergraphs (and trend lines)	

Whichever presentation method is chosen it should be:

* **clear and easy to understand**
* **as simple as possible**
* **able to get the message across to the reader.**

7.2 Graphical methods

1 ARITHMETIC GRAPHS

Arithmetic graphs are used with linear scales where divisions on the axes represent equal amounts of the data; one division is always 10, 25 or 100 units, or whatever is appropriate. Try to avoid awkward scales, e.g. where 1 cm on the graph paper represents a number which is not only difficult to plot but which is also difficult for the reader to 'read back'. It is better to use numbers such as 1, 5, 10, 20, 25, 50, 100 etc. as the basis of the scale divisions.

The construction of an arithmetic graph is as follows:

a   **Decide which axis is to be used for each set of data. Usually the independent variable is plotted on the horizontal axis and the dependent variable on the vertical axis. Think about the two variables and decide which one influences the other, e.g. if you were plotting a graph of employment changes over a period of years it is the employment which is influenced by time; therefore the employment figures would be plotted on the vertical axis (dependent) and the time periods on the horizontal axis (independent). The time periods are 'fixed' i.e. they are independent.**

b   **Choose a scale for each axis which allows you to plot the full range of data for each variable.**

c   **Label the axes clearly and indicate the scale divisions.**

d   **Plot the data and draw a line between the points.**

e   **Leave space for a title and key if you are plotting more than one line on the same axes. This can be done by using different symbols for the plots, or by using dotted and solid lines between the plots.**

It is possible to show two sets of data on the same graph; the left hand vertical axis can be used for one scale with the right hand vertical axis being used for a different scale. The different scales make it difficult to compare actual changes but they can give a useful visual impression of the connection between two sets of data.

For example, birth rate and death rate could be scaled on one axis with total population on the other. Demographic changes could then be seen in relation to the overall change in population total.

Arithmetic graphs are appropriate when you want to show absolute changes in data (see Figure 7.1).

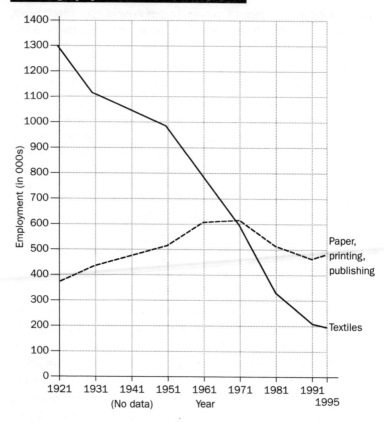

Figure 7.1 A graph to show changes in employment in textiles and paper, printing and publishing (1921–1995)

2 LOGARITHMIC GRAPHS

Logarithmic scales are divided into a number of cycles each representing a tenfold increase in the range of values. If the first cycle ranges from 1 to 10, the second extends from 10 to 100, the third from 100 to 1000 and so on. Unlike arithmetic graphs each division on the scale does not represent an equal range of data.

The intervals on the log scale are all exponents of 10; the scale can start at any exponent of 10, for example, 0.001 to 0.01, 0.01 to 0.1, 0.1 to 1.0 for very small values; or 1,000 to 10,000, 10,000 to 100,000, 100,000 to 1 million for large values. The total number of cycles needed depends upon the full range of data to be plotted.

Zero cannot be plotted on logarithmic scales, nor can positive and negative values be shown on the same axis; therefore log scales are of no use if you want to show this type of data.

Logarithmic graphs are good for showing the rate of change rather than absolute change; a steeper line indicates a faster rate of change. They also allow a wider range of data to be displayed than on a similar sized piece of arithmetic graph paper.

For line graphs we usually use semi-logarithmic graph paper on which one axis consists of one or more log cycles and the other is a linear or arithmetic scale. If the rate of change is increasing at a constant proportional rate i.e. the population doubles in each time period, this will appear as a straight line on semi-log paper where time periods are plotted on the linear scale and population (dependent variable) is plotted on the logarithmic scale.

The construction of a logarithmic graph is as follows:

a   **Select a piece of graph paper which has sufficient cycles to cover the full range of data which is to be plotted, remember that the cycles must begin and end at an exponent of 10.**

b   **Allocate axes in the same way as for arithmetic graphs; decide which data set needs to be plotted on the log scale; for graphs of changes over time, time period would always be plotted on the arithmetic axis.**

c   **On your log axis, label the start point of each cycle and intervening points; label the linear axis as you would for an arithmetic graph.**

d   **Plot the points for the data sets and draw a line between the points.**

e   **Label the axes; title the graph and provide a key where necessary.**

3 CUMULATIVE FREQUENCY CURVES

These can be drawn using either arithmetic or logarithmic axes. The data for each class in a distribution is converted into a percentage and each percentage figure is added successively.

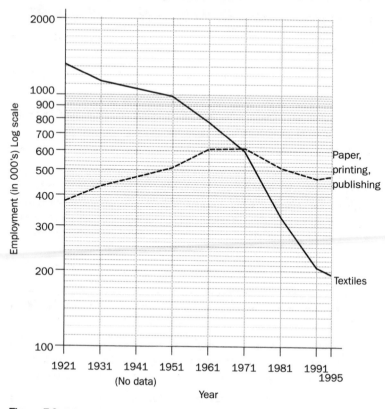

Figure 7.2 A logarithmic graph to show employment change in textiles and paper, printing and publishing (using the same data as for the arithmetic graph (Figure 7.1))

For example, if 5% of the population is under five years,

4% is 5–9, 6% is 10–14

then cumulatively   under 5 = 5%

5–9 = 5% + 4% = 9%

10–14 = 5% + 4% + 6% = 15%

This would be continued until all class groups have been included when the total cumulative percentage would be 100 per cent.

The construction of a cumulative frequency graph is as follows:

a **Arrange the data into cumulative percentage form.**

b **Draw and label the vertical axis 'cumulative percentage' with a scale from 0 to 100 per cent.**

c **The horizontal axis should be labelled with a range of values as determined by the class groups you are using.**

d **Plot individual points using the co-ordinates for cumulative percentage and each class group in turn.**

e **Join the points with a straight line to show the frequency distribution.**

Projects in Practice

Lisa has collected samples from material which has been deposited on the floor of a glaciated valley. She analyses the samples by sieving and weighing in order to identify the proportion (by weight) of each grain size. The results from two samples are:

Grain size (mm)	Sample A	Cumulative %	Sample B	Cumulative %
0.001	27.0	27.0	0.7	0.7
0.004	30.4	57.4	1.1	1.8
0.063	11.0	68.4	5.5	7.3
0.25	12.1	80.5	7.3	14.6
0.5	15.7	96.2	52.6	67.2
2.0	1.8	98.0	11.1	78.3
4.0	1.8	99.8	11.5	89.8
64.0	0.2	100.0	10.2	100.0

Because of the wide range of values for grain size, Lisa decided to use semi-logarithmic graph paper; the vertical (arithmetic) scale was numbered 0–100 and labelled 'cumulative percent', and the horizontal (logarithmic) axis was labelled in cycles to cover the range of grain sizes. The first cycle started at 0.001; successive cycles began at 0.01, 0.1, 1.0, 10.0 (ending at 100). This allowed any value between 0.001 and 100 mm to be plotted. The largest grain size was 64 mm.

Continued

The two samples display very different characteristics; A consists of a wider range of particle sizes but with very little greater than 5 mm. In contrast, B has a much smaller range of particle sizes, mainly between 0.25 and 2 mm. This would suggest that the two samples had been deposited in different ways; the degree of sorting in B could indicate that this deposit had been laid down under fluvio-glacial conditions with sorting by water, whereas the wider range of sizes in A could suggest that this is an unstratified deposit laid down under ice.

The median value (the 50th percentile) can be identified by finding the value of the grain size at the cumulative percentage 50 line; for A it is 0.0029 mm, for B it is 0.39 mm.

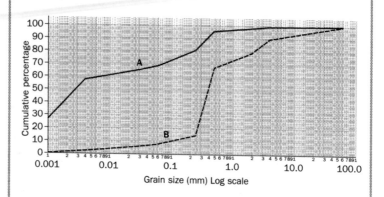

Figure 7.3 Cumulative percentage grain size for samples A and B

4 LORENZ CURVE

This is a particular type of cumulative frequency graph which can be used to measure, or illustrate, the extent to which a geographical distribution is even or concentrated. For example, the percentage of working population in a particular industry in each region of the UK could be compared with the percentage of national employment in each region. This would indicate the extent to which employment in that industry was distributed in relation to the workforce nationally. The regions must be placed in rank order according to the importance of that industry. Both sets of data are arranged in cumulative percentage form, but in rank order with the largest category first. This provides some measure of the degree of spatial concentration; this can be related to the Location Quotient (see Chapter 8) which represents geographical concentration as a numerical value.

For other distributions in which you do not need to compare with a national or regional distribution the cumulative percentage for one set of data can simply be plotted in rank order. In this case, the vertical axis is labelled 'cumulative percentage' and is scaled 0–100, the horizontal axis is labelled 'Rank order'

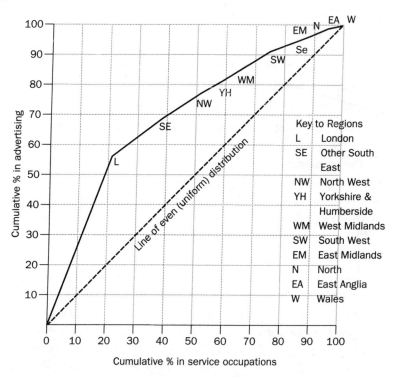

Figure 7.4 A Lorenz curve to show the cumulative percentage of service workers in advertising in the standard regions of the UK in relation to the cumulative percentage of employment in all service occupations

and is scaled to cover the total number of items in the data set (for example 10 categories of employment would require 10 rank orders).

This allows comparison with other distributions and the line of perfect regularity (line of even distribution). This could be used in assessing the extent to which employment is concentrated into particular occupations, or the degree to which a country depends upon certain types of energy.

The construction of Lorenz curve is as follows:

a   Draw up axes which are scaled 0–100 per cent with zero as the origin for both axes. If plotting a distribution against 'rank order', then the horizontal axis would be labelled accordingly and would be numbered from 1 (e.g. largest category of employment) to whatever rank was required (i.e. the total number of categories, *n*). In this case leave one division on the horizontal axis before labelling 'rank 1'.

b   Draw the line of 'perfect distribution' (even distribution) from the origin (0,0) to the point at which both variables are 100; or the point where all the categories have been included (co-ordinate 100, rank *n*).

c   Arrange the data in order of size (rank order).

d   Calculate each as a percentage of the total distribution.

e   Add the percentages cumulatively, as shown for cumulative frequency curves.

f   Plot the points and join the dots with straight line segments.

If your classes are in proper rank order then the gradient of your lines should get progressively shallower because you have added successively smaller values to your cumulative percentage figures as you have moved down the ranks.

How to interpret the curve:

a   If your line is fairly close to the 'line of even distribution' then the data reveals a fairly even spread with no major concentrations.

b   If the Lorenz curve 'bulges' away from the line of even distribution then this shows some degree of unevenness in the pattern. The further the line is away from the diagonal line then the greater the degree of concentration into a few categories. The greater the angle between the diagonal line (even distribution) and the plotted Lorenz curve the greater the concentration into the leading regions or categories.

Projects in Practice

As part of a study of the heritage industry in the UK, Gareth has researched some data about museum and art gallery visitors in terms of their social class. In order to assess the relevance of social class as a factor in the development of heritage activities he decides to use a Lorenz curve to compare the data with the class distribution of the UK population, as shown below.

Class	% Museum/ gallery visitors	% UK adult population
A/B	33	19
C1	27	23
C2	24	28
D	11	17
E	5	13

The 'visitors' data is already in rank order; Gareth calculates the cumulative percentage values for both sets of data and plots these as co-ordinates for class A/B through to E i.e.

A/B co-ordinates	33, 19
C1	60, 42
C2	84, 70
D	95, 87
E	100, 100

The graph shows some degree of unevenness with a higher concentration of visitors from the social classes A/B and C1, with 60 per cent of visits compared with 42 per cent of base population. Groups D and E are both 'under-represented'. Data for other tourist activities, such as steam railways or country parks, could be plotted on the same axes for comparison.

Continued

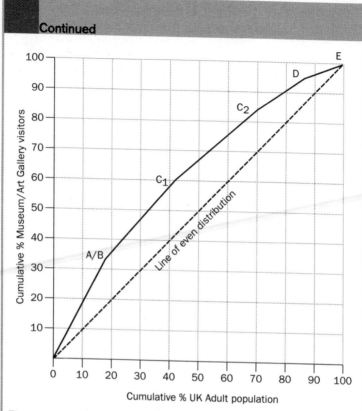

Figure 7.5 Lorenz curve: visitors to museums and galleries against UK adult population by social class.

5 POLAR GRAPHS

These graphs are particularly useful when one variable is a directional feature, e.g. a wind rose diagram which shows both the direction and the frequency of winds. The circumference represents the compass directions and the radius can be scaled to show the percentage of time that winds blow from each direction.

They can also be used when one variable is a recurrent feature such as a time period of 24 hours, or an annual cycle of activity. This could illustrate traffic flows or pedestrian flows over a period of time during the day, or monthly output figures.

Figure 7.6 shows the number of pedestrians passing a given point (e.g. the PLVP in centre of the CBD) per minute recorded between 0600 and midnight. The circular scale is used for hours during the recording period and the radial scale represents number of people.

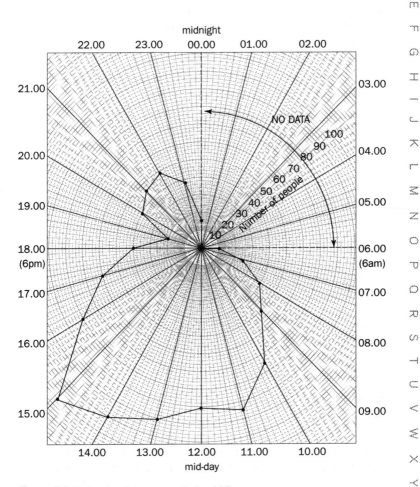

Figure 7.6 Pedestrian flow patterns in the CBD

The construction of a polar graph is as follows:

a   **Label and scale the circumference according to the type of data. If the graph is being used to illustrate 'directional' data then the cardinal points of the compass can be placed in their normal position with North in the vertical position from the centre point. Working clockwise, at intervals of 45° label the points NE, E, SE, S, SW,W, NW and back to N. Other bearings, as required, can be inserted at intervals of 22.5°. For 'daily' data, each hour is represented by 360°/24, i.e. 15° around the circumference. For 'annual' data, each month is represented by 360°/12, i.e. 30° around the circumference.**

b   **Select a suitable scale for the radius. Measure the scale outwards from the centre as a series of concentric circles.**

c   **Plot the points which, depending on the data, may be left as a pattern of dots, joined with lines around the circle, or plotted as bars from the centre of the circle.**

Such graphs should only be used for specific types of data; the scale around the circumference can only be used for a variable which is continuous, such as a repeating time sequence or the points of a compass. Because the axes are so unusual, and they tend to visually distort the spread of values, they are often difficult to interpret.

Projects in Practice

Lucy is investigating the orientation of corries around a series of glaciated peaks. She has collected data relating to the orientation of each corrie by measuring the direction in which the backwall of the corrie is facing. These values are recorded as a number between 0° (due North) and 359°.

She also determined the height of each corrie from an OS map. This was taken as the height (in metres) of the corrie lip where the stream emerged from the rock basin.

In order to display the data, and to identify any patterns in the orientation of corries, Lucy uses a polar graph.

The circumference is scaled as compass bearings from 0° clockwise to 360° (0°) i.e. a full circle. For the radial scale, Lucy decides to plot the highest values near the centre. She

Continued

works outwards from 750 m (which accommodates the highest corrie at 737 m) at intervals of 50 metres. From the graph Lucy is able to identify the main sectors in which corries develop and the range of heights at which corries form in different sectors. She could then follow this up with some statistical analysis, such as chi square, in order to see if there is any significance in this pattern. This could then provide further lines for investigation.

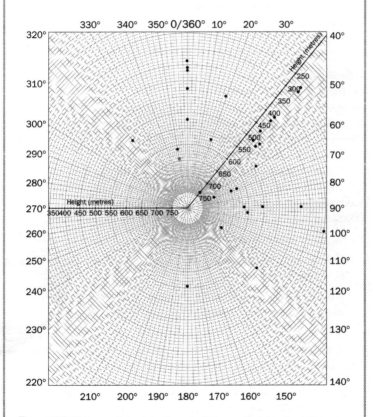

Figure 7.7 Polar graph to show the orientation and height of corries.

Continued

The graph shows that the distribution of corries is not even; there are more corries in the quadrant between 0 and 90 degrees and these corries are found over a wider height range than corries between 90 and 270 degrees. This pattern could lead on to further investigation concerning the factors which encourage the accumulation of ice and the processes which lead to corrie formation.

6 PIE GRAPHS (OR PIE CHARTS, DIVIDED CIRCLES)

In these graphs the area of the circle is divided into segments according to the share of the total value represented by that segment. Each segment is calculated as a number of degrees according to its proportion of the whole circle. These are visually quite effective; it is easy to see the relative contribution of individual segments to the whole. However, they are difficult to read, it is hard to assess percentages accurately from the pie chart especially if individual 'slices' are small or there are a large number of segments. Comparing one pie graph with another at anything beyond a superficial level is difficult. Small segments, less than 5° are difficult to see; if necessary group some sections together to avoid too many, or too small divisions.

The construction of a pie graph is as follows:

a **Draw a circle; this should be large enough to allow a protractor to be used easily.**

b **If the data is in percentage form then multiply each percentage value by 3.6 to convert it to degrees (a full circle is 360°, therefore 1 per cent is 360/100 or 3.6°).**

If the data is in raw form, i.e. absolute values, then, for each component of the data, use the formula:

$$\frac{X}{\text{total}} \times 360$$ where X is the value of the individual component
total is the sum of all the components

The answer is the number of degrees for that component.

c   Measure the angle for each segment around the circle starting at the vertical position (12 o'clock on the face of the circle). Normally the largest segment is drawn first, with progressively smaller segments as you work clockwise. Join the mark on the circumference to the centre of the circle.

d   Shade, or label, each segment and include a key if required.

e   Annotations giving the raw data are often added around the edge of the pie graph; although this makes it easier to read back information, to some extent it undermines the method of presentation; it does suggest that the presenter realises that the technique has limitations, it might also suggest that you should have chosen a more appropriate technique!

7 BAR GRAPHS

Bar graphs (or bar charts) consist of a series of vertical bars or columns rising from a horizontal axis. The height of the bar is proportional to the quantity represented. The vertical scale can be used to represent percentage values or absolute data. The horizontal scale can be used for any discrete or categorical data. A bar graph could be used to show, for example, the total number of people employed for each year over a period of years. It could also be used to show different categories in one time period, for example, the percentage of the shoppers who arrive at a shopping centre by different forms of transport; car, bus, train, tram, cycle, walk.

Other variations consist of:

●   **multiple bar charts, where several bars for each year can be plotted against a common vertical axis**

●   **divided bar charts, where a rectangular bar representing 100 per cent is divided up into its constituent parts for each sub-division of the data; for example, employment sub-divided into primary, secondary and tertiary percentages**

●   **the population pyramid which shows the age/sex structure of population by plotting the percentage of population (male and female) in each five year age cohort.**

Bar charts are effective because they are so commonly used and therefore they tend to be easily understood. They show relative magnitudes very effectively and by having a scale which passes through zero they can be used to show positive and negative values, such as profit and loss, on the same graph. It is also relatively easy to read back data from a bar chart because the height of the bar can be easily compared with the vertical scale. This makes them much more versatile and useful than pie charts. Because of this they may be over-complicated by the presenter including too many multiple bars or by using two different scales on the vertical axes.

Do not be tempted to make the simple bar graph more exciting; its greatest asset is its simplicity and clarity of presentation!

The construction of a bar graph is as follows:

a **Draw up appropriate axes with categories or time periods on the horizontal axis and frequencies as percentages or absolute values on the vertical axis.**

b **Plot the values for each category or time period as bars with spaces between each bar. Bars should be of equal width with the height proportional to its value according to your scale. It is perfectly acceptable to present horizontal rather than vertical bars; you simply switch the axes; it depends on what you think is the clearest way of displaying your data. This can work effectively when you are trying to show differences for a number of items from year to year.**

c **Label both axes so that it is clear what they represent.**

8 PROPORTIONAL SYMBOLS

In this presentational method a symbol is drawn proportional to the value it represents; this symbol could be a bar (as described in section 7), a circle, a square, a cube or a sphere. They may be used independently or as plots on a base map to show spatial differences. When used as a cartographic method, you need to exercise great care; because of the varying size of each plot it is often difficult to foresee the end product and very careful planning and placing of symbols is needed to avoid overlap and confusion.

Remember, the key point in any presentation method is its clarity in presenting the data.

The most commonly used proportional symbols are the bar and the circle. The bar would be constructed as shown on page 87.

Proportional circles:

● **the area of the circle represents the total value of the data**

as the area of a circle is given by the formula πr^2

if you wished to represent a total value of V units, then $V = \pi r^2$

the radius of the circle required to represent a value V would be:
$r = \sqrt{(V/\pi)}$

The construction of a proportional circle is as follows:

a   **Substitute the value which you wish to represent into the formula $r = \sqrt{(V/\pi)}$. Take the value for π as 3.142.**

b   **Draw a circle of radius r.**

NB Often the calculated value for r is too large for you to fit the circle on to a standard size of graph paper. It will therefore be necessary for you to adjust the scale until the circle is a convenient size.

It is important that you are able to state the scale of your proportional circle.

Whilst it is easy to reduce the calculated value of r until the circle fits the paper, this can make it difficult for the presenter to appreciate what has happened to the scale.

If you halve the value of r, you will quadruple the scale; if you divide the value of r by three then you will increase the scale by a factor of 9.

To avoid this problem it may be easier to think about the problem of the scale at the outset.

A proportional circle has a scale where the area represents the required value; therefore the scale is in the form of:

● **1 square unit on the graph paper represents x units of data.**

The first step should be to assume a scale; let us say 1 sq cm represents 1 million people.

If a country had a population of 60 million then using the formula

$$r_{60} = \sqrt{(V/\pi)} ,$$
$$r = \sqrt{(60/3.142)}$$

   $= 4.37$ cm A circle of radius 4.4 cm would be drawn.

If one of the countries that you wished to represent had a population of 600 million then the radius required would be

$$r_{600} = \sqrt{(600/3.142)} = 13.8 \text{ cm.}$$

This would be difficult to fit on the graph paper.

Therefore the scale would need to be adjusted.

In order for you to retain control of the scale, simply assume another scale and work it through until you arrive at values for the radius which will fit.

For example, if you now assume that 1 cm^2 represents 2 million people then for a value of 600 million to be represented you will need to enclose an area of 600/2 i.e. 300 cm^2.

The radius would be

$$r_{300} = \sqrt{(300/3.142)} = 9.8 \text{ cm}.$$

A general formula can be expressed for this adjustment.

$$r = \sqrt{\frac{(V \div n)}{\pi}}$$ where n is the scale adjustment you are using.

If 1 cm^2 = 1 unit (1 million in this case) then $n = 1$

If 1 cm^2 = 5 units (i.e. 5 million) then $n = 5$.

By adopting this approach you keep control of the scale and can state the scale clearly for any diagram which you present.

A similar process can be applied to the less commonly used symbols.

For **proportional spheres**, the size is calculated as the area of the square; therefore the length of the side of the square is found by using the formula $L = \sqrt{V}$ (the scale is expressed in the form 1 square unit represents x units).

Adjustments to the scale can be made as for proportional circles.

When using **proportional spheres** it is the volume that represents the total value. Although this enables them to represent a much wider range of values it makes them difficult to draw and interpret.

The radius of the sphere is given by the formula $r = \sqrt[3]{\frac{3V}{4\pi}}$ i.e. multiply the value by 3 ($V \times 3$) and divide by 4π (12.56). Then find the cube root of this value.

Proportional cubes are also solid objects and their volume also represents the value you wish to present. Although easier to draw than proportional spheres they are also difficult to interpret when comparing volumes of a set of cubes placed on a base map.

The length of the cube's side is given by the formula $L = \sqrt[3]{V}$ i.e. the cube root of the value.

Projects in Practice

Helen is comparing the levels of investment in the research, development and design of different forms of energy production and conservation measures in a number of countries in Europe. She decides to use proportional divided circles to illustrate the differences between countries. The data for Sweden and UK is as follows:

Spending on RD&D (millions of $US) 1990

Sweden	Energy type	UK
7.3	Fossil fuels	28.3
14.9	Nuclear	273.2
18.1	Renewables	21.5
23.7	Conservation	36.8
64.0	Total	359.8

In order to work out the scale of the circles she initially assumes that 1 cm2 represents $1 million.

Therefore, for the larger circle, the UK, the radius of the circle would be:

$$\text{radius (UK)} = r_{UK} = \sqrt{\frac{359.8}{\pi}} = 10.7 \text{ cm}$$

For Sweden the radius would be 4.5 cm.

The circle for the UK would be too large for the space available; Helen adjusted the size of the circles by assuming a different scale:

If 1 cm^2 represents $2 million then the radius for the UK would be

$$r_{UK} = \sqrt{\frac{359.8/2}{3.142}} = 7.6 \text{ cm}$$

and for Sweden $r = 3.2$ cm. Circles of this size are suitable.

The circles are sub-divided according to the percentage of investment in each energy type; 1 per cent is represented by 3.6°.

Continued

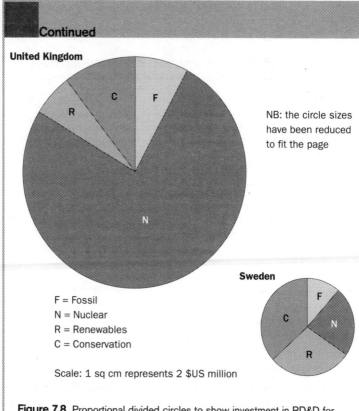

United Kingdom

NB: the circle sizes have been reduced to fit the page

Sweden

F = Fossil
N = Nuclear
R = Renewables
C = Conservation

Scale: 1 sq cm represents 2 $US million

Figure 7.8 Proportional divided circles to show investment in RD&D for different energy types. ($US million)

9 LONG AND CROSS SECTIONS

These can be used to describe and compare the shape of the landscape in different areas. Although long sections are most commonly applied to river profiles, cross sections could be used to describe any landform feature. They are, effectively, a graph displaying height against a horizontal scale of distance. The horizontal axis, or base line of the section, is simply taken from one point to another across the map; it is therefore normally at the same scale as the map. If the length of the section is not suitable for your purpose, for example if you wish to make it larger to accommodate some information which you need to add or the base line is too long to fit the graph paper available, then the scale can be adjusted.

If you are using a 1:50 000 scale map then 1 cm on the map represents 50 000 cm on the land i.e. 500 metres. (For a 1:25 000 scale, then, 1 cm represents 250 m.) These would be the normal scales for your base line, but could be adjusted to suit your purpose.

It would be unusual to use the same scale on the vertical axis because a scale of 1 cm to 500 m height would produce a profile which would show little variation for many of the landforms which you are likely to investigate. It is usually necessary to adopt a larger scale for the vertical axis; however some care must be taken not to use a scale which presents quite small features as major or precipitous landforms! A scale of 1cm represents 100 m would be acceptable for most landform sections. Even at this scale, for a 1:50 000 map, the vertical scale is exaggerated. The degree of this exaggeration can be calculated and stated as part of the section.

By comparing the horizontal scale: 1:50 000 or 1 cm = 500 m with the vertical scale, 1 cm = 100 m it is clear that the vertical scale is five times larger than the horizontal scale i.e. 5 cm would represent 500 m on the vertical scale: the vertical exaggeration is 5.

The construction of a cross section is as follows:

a   **Use a piece of paper with a straight edge and place this across the map between the two points which form the ends of your required section. Mark the ends of this base line as A and B and identify these points using a six figure grid reference.**

b   **On the paper, mark the points where each contour line cuts the straight edge and note the height of each of these contour lines.**

c   **Using arithmetic graph paper, draw a base line of length A–B and a vertical axis with an appropriate scale.**

d   **Transfer the information from your recording paper to the graph paper by using the contour intersection points and the height of each contour as co-ordinates for your plots.**

e   **Join the plots with a smooth line; use the map to interpret the shape of the relief between the fixed contour plots.**

A B C D E F G H I J K L M N O P Q R S T U V W X Y Z

f **Calculate the vertical exaggeration and state this below your section, which should be fully labelled to indicate location and scale.**

For a long section, the same procedure is used but the base line, i.e. the straight edge of your recording paper, has to follow the course of the river. You should begin at the relevant point, usually the source of the river, and mark the points where contours of known height cut across the river channel. The end point of the profile may be a major confluence, a lake or the estuary; this will depend upon the scale and purpose of the study.

Move your recording paper around the bends in the river channel so that the base line represents the distance along the channel.

The long profile is constructed as above, however the scale may need to be adjusted. The simplest way of doing this is to measure each contour intersection from your start point and adjust all measurements by the same scaling, for example, reduce them all to one half of the original distance.

10 DISPERSION GRAPHS/BOX AND WHISKER PLOTS

These graphs are used to display the main patterns in the distribution of data. A dispersion graph shows each value plotted as an individual point against a vertical scale; it shows the total spread, or range, of the data and the distribution of each piece of data within that range. It enables some comparison of the degree of bunching of the data. Box and whisker plots provide more detail on the range and spread of data. They indicate the position of the median, the upper quartile, the lower quartile (inter-quartile range), and the highest and lowest values (the range).

These measures of central tendency (range, median, quartiles etc.) are explained on pages 113–116.

The construction of dispersion graphs/box and whisker plots is as follows:

a **Draw a graph with categories of data on the horizontal scale and values or frequencies on the vertical scale.**

b **Scale the vertical axis to accommodate the full range of your values or frequencies.**

c **For each category of data, plot each value vertically against the scale. If two, or more, pieces of data have the same value then plot them at the same height next to each other.**

d   To present a box and whisker plot for the same data then calculate the median, the upper and the lower quartile and plot these as short bars running parallel to the horizontal axis.

e   Draw vertical lines from the upper quartile value to the lower quartile value to 'box' this spread of data.

f   Mark the highest and lowest values by drawing lines parallel to the horizontal axis, as above.

g   Join these to the 'box'. These 'whiskers' show the overall spread (range) of the data.

The 'box' shows the spread of the middle 50 per cent of the data values i.e. the inter-quartile range.

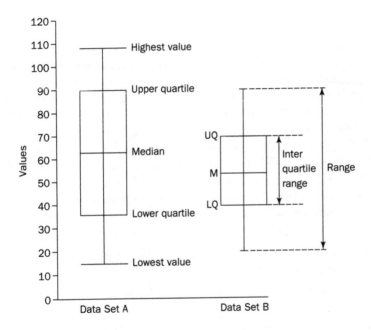

Figure 7.9 Box and whisker plots

11 HISTOGRAMS

Histograms are used to display, and analyse, the frequency distribution of data. Bars are drawn to indicate the frequency of each class of data, but unlike bar graphs, it is the area of the bars rather than their height which shows the distribution of the data. They are commonly used to analyse the distribution of values for continuous data within a measured range of possible values; for example rainfall totals, farm size, income levels or distances from a fixed point such as the centre of a CBD. They are used to simplify and clarify data which may be easier to analyse when placed into groups, or classes, rather than individual data. A large amount of data can be reduced to more manageable proportions which may allow the identification of underlying trends.

Grouping the data into appropriate classes can be quite difficult and requires some thought and preliminary consideration of the data. The method which you select should reflect the nature of the data which you are displaying. Your main aim should be to illustrate the important differences between classes whilst achieving minimum variation within each class.

Although it is difficult to provide any fixed rules which control the choice of classes for a particular data set, there are some general principles which you should follow in trying to decide upon the number of classes and the range of values in each class.

a   **The number of classes used will depend upon the total number of individual values which you have collected as the data set. If you have too many then there may be insufficient variation between the classes, or some classes may not contain any of the individual values; although 'empty' classes may be unavoidable when the data has a very wide range. Equally, if you have too few then the range of values within a class may be too large; this could lead to loss of detail and make it difficult to identify any trends. As a general guide, the maximum number of classes selected can be determined using the formula:**

Number of classes = 5 × logarithm of the total number of items in the data set.

So, if you had collected data about the size of 120 farms in an area: the maximum number of classes

$$= 5 \times \log 120$$

$$= 5 \times 2.08$$

$$= 10.4 \text{ or } 10 \text{ to nearest whole number.}$$

Fewer classes could be used if that would produce a histogram which was easier to interpret. The formula should only be used to gain some idea of the maximum number of classes – if you only had 25 data items the application of the formula would indicate that the maximum number of classes should be $5 \times \log 25 = 7$. In practice you might decide that four or five classes were more suitable when you only had such a small number of items in the data set.

b   The range of values in each class (the class interval) will obviously be influenced by the number of classes which you decide to employ. The simplest form of histogram which can be drawn, and interpreted, is one based upon classes with a fixed interval, i.e. each class includes an identical spread of data, e.g. 0–10, 10–20, 20–30 etc. The class interval and the relevant class boundaries can be calculated using the formula:

$$\text{class interval} = \frac{\text{range of values (highest-lowest)}}{\text{number of classes}}$$

If the data ranges from 4 to 98 and you require 4 classes, the class interval would be

$$= \frac{98 - 4}{4} = 23.5$$

It may be more convenient to extend the range from 0 to 100 and adopt a class interval of 25 in this case.

The class boundaries must be carefully defined in order to avoid any problems, and possible confusion, in the allocation of individual data values to classes.

For example, if you adopted a class interval of 25; the class boundaries would be 25, 50, 75, 100 and the successive classes would cover the following data ranges: 0–25, 25–50, 50–75, 75–100.

If any of the individual values were 25, 50 or 75 then they could be placed in two classes; this clearly would not be appropriate and the class boundaries would need to be adjusted to: 0–24.9, 25–49.9, 50–74.9, 75–100.

These are now continuous classes with no overlap.

Careful selection of the number of classes and the class interval can avoid such problems.

It is possible to draw a histogram with different class intervals e.g. 0–4.9, 5–9.9, 10–19.9, 20–29.9.

Here the first two classes have a data range which is half the value of the third and fourth class. If this was selected as a suitable class boundary system then you must remember that it is the area of the bars that describes the frequency of the distribution. The height of the bars for classes 3 and 4 would need to be reduced to a half of the measured frequency in relation to bars for class 1 and 2. This would then present the area of the bar correctly.

A simple method of identifying possible class boundaries is to plot the data in the form of a dispersion graph which shows individual values. This may enable you to locate natural breaks within the distribution of data; the choice of such breaks is likely to be subjective although it may help you to determine the most appropriate number of classes for the data.

The process of sub-dividing the data into groups is not the main purpose of the exercise; it is not an end in itself!

The decision about the number of classes and the class interval should be influenced by the type of data which you are presenting and the purpose to which it is being put. You need to decide exactly what it is that you are trying to illustrate or analyse.

The construction of a histogram is as follows:

a   Arrange your data into classes according to the class boundaries which you adopt. Count the number of individual values which lie within each class; this is the frequency in each class.

b   Using arithmetic graph paper, draw and label two axes; a horizontal axis of sufficient length to accommodate the number of classes and a vertical axis scaled to cover the value of the largest class frequency.

c   Plot the frequency value for each class as a bar; these should be drawn as contiguous bars i.e. touching each other with no gaps between.

d   If all of your classes cover an equal range of values then your divisions on the horizontal scale should be uniform; if classes cover a different range of values then adjust the width of the class divisions accordingly.

e   Ensure that your histogram is given a title and is fully labelled.

f   You may wish to identify and label the modal class i.e. the class size which has the highest frequency.

In commenting on data distributions it is often useful to compare the plotted distribution with some general patterns. If your histogram has a modal class in the middle of the class range with progressively smaller bars on either side then it is similar to the 'normal distribution'. If the modal class lies in the lower class sizes then the distribution can be described as having a 'positive skew'; if the modal class lies in the higher class sizes then the distribution has a 'negative skew'.

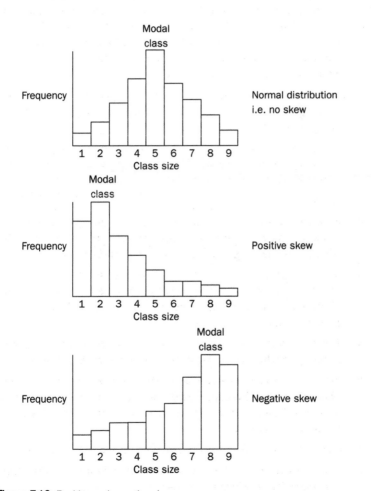

Figure 7.10 Positive and negative skew

Projects in Practice

Peter is undertaking an investigation into aspects of agricultural change; he has collected data relating to the size of individual farms in one area at two different dates. In order to identify and summarise any differences between the two periods he decides to draw a histogram for each set of data. These are drawn up using the same class intervals and class boundaries to allow direct comparison of the farm size distribution.

He has data for 49 farms in 1840 and 33 farms in 1960.

Using the formula to determine the maximum number of classes:

Number of classes = 5 log 49 = 5 x 1.69 = 8.5

He decides to use 8 classes with a class interval of 25 but as the areas are to 1 decimal place he uses a tally chart of classes from 0–24.9, 25–49.9 etc.

	1840	**1960**
0–24.9	21	7
25–49.9	13	5
50–74.9	8	6
75–99.9	1	4
100–124.9	4	4
125–149.9	2	1
150–174.9	–	5
175–200	–	1

The histogram for 1840 shows a positive skew with the modal class 0–24.9. The histogram for 1960 shows a more even distribution with a higher proportion of larger farms although the modal class remains at 0–24.9.

Continued

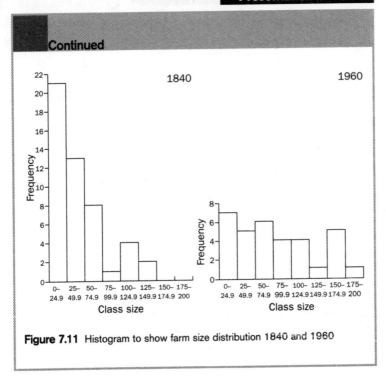

Figure 7.11 Histogram to show farm size distribution 1840 and 1960

12 TRIANGULAR GRAPHS

Triangular graphs are plotted on specialised graph paper which is drawn in the form of an equilateral triangle. Each side of the triangle is used for one variable. It is a useful method for examining the varying proportions of three related sets of data, for example the percentage employment in primary, secondary or tertiary industry or the classification of soils in relation to the percentage sand, silt or clay in the soil fabric.

The data must be in percentage form and the three percentage values must add up to 100 per cent; this form of graph paper can only be used when there are three variables which contribute to a complete data set; it cannot be used to represent absolute values, and so raw data must be converted to percentage form.

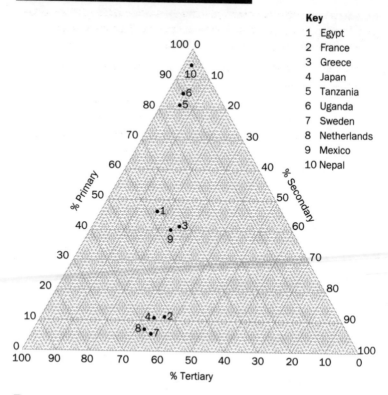

Key

1 Egypt
2 France
3 Greece
4 Japan
5 Tanzania
6 Uganda
7 Sweden
8 Netherlands
9 Mexico
10 Nepal

Figure 7.12 Countries in different economic stages; employment in primary, secondary and tertiary industry

Triangular graphs are particularly useful in showing the relative importance of each of the three sets of data and they give a clear visual impression of the dominant variable. Any sub-grouping of points in the form of clusters in different areas of the graph may enable classification of the items according to the extent of dominance of one variable or the degree to which the proportions are balanced.

The construction of a triangular graph is as follows:

a   Using triangular graph paper, scale each side of the triangle from 0 to 100 per cent; begin at one apex and label the divisions from 0 up to 100 at the next apex; repeat this labelling working round the triangle.

It does not matter whether you move around the sides in a clockwise or anti-clockwise direction, but be consistent. Each apex should have one 0 per cent and one 100 per cent label.

b **Make sure that your data is in percentage form and that the three values add up to 100.**

c **Plot the individual points at the % A, % B, % C intersection. As a check: if you locate the point where the % for A and the % for B intersect then the % for C should follow automatically as the three values add up to 100. If this does not occur then you are using the lines on the graph paper incorrectly.**

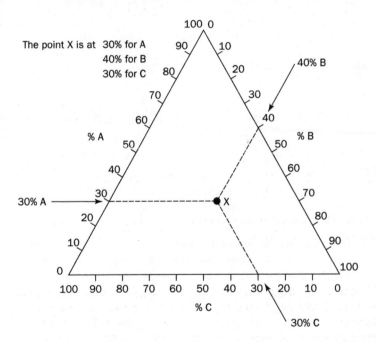

Figure 7.13 Constructing a triangular graph

13 SCATTERGRAPHS

Scattergraphs are often used in the initial stage of data analysis to investigate the relationship between two sets of data. Although they are used as a form of presentation they are particularly useful in identifying any patterns or trends in the relationship which might suggest the need for further inquiry.

By inserting a trend line on the graph the general relationship can be observed. The positioning of a trend line through an array of points is subjective, i.e. different researchers may have different perceptions of the trend. Therefore a more objective test of the extent of the relationship or correlation between the two variables is often needed. A commonly used correlation test is the Spearman Rank correlation coefficient which is described on page 119.

Scattergraphs can be plotted on arithmetic, logarithmic or semi-logarithmic graph paper. It is important that there should be some logical geographical reason for investigating the relationship between the two variables otherwise the observed trend may be purely coincidental or 'spurious'. There is usually some suggestion of a cause and effect between the two variables; one could be seen as the independent variable and the other as the dependent variable (as discussed for arithmetic graphs on page 72).

The direction of the trend line can be studied to gain a general impression of the relationship:

* **if as one variable increases the other increases; the trend is positive; the line goes from bottom left to top right with a positive gradient – a direct relationship**

* **if as one variable increases the other decreases; the trend is negative; the line goes from top left to bottom right with a negative gradient – an indirect relationship**

* **if there is no recognisable pattern; this suggests that there is no relationship.**

The closer the points are to the trend line then the greater the degree of relationship, but this can only really be assessed with any degree of reliability with a statistical test.

Individual points which lie some distance away from the trend line are classed as residuals (or anomalies); above the line they are positive residuals; below the line they are termed negative residuals. Any such points which do not appear to be following the perceived trend are often worthy of further study to gain greater insight into the external factors which may influence the two selected variables.

The construction of a scattergraph is as follows:

a   Draw two axes with the independent variable on the horizontal axis and the dependent variable on the vertical axis.

b   Scale and label each axis to accommodate the range of values for each variable.

c   Plot each point using the values for the x and y axes as co-ordinates. It may be useful to identify each point by a reference number or letter to assist you in your comments or identification of residuals.

d   Insert a 'best fit line' by eye, that is, try to place a trend line through the middle of the spread of points taking account of the low and high values.

A more accurate method can be used to locate the trend line. This involves the use of 'semi-averages' to generate a regression line which summarises the characteristics of the scattergraph.

The procedure is as follows:

a   Calculate the mean value for the data on the independent axis (x) and the mean value for the data on the dependent axis (y).

b   Use these two values as co-ordinates to plot a point on the scattergraph labelled A.

c   Using only the values of x which are higher than the mean ($\bar{x}$), calculate the upper semi-average; add the relevant values and divide by the number of items. Repeat this process for the values of y which are higher than the mean ($\bar{y}$). Use these values as co-ordinates to plot a point on the scattergraph labelled B.

d   Calculate the lower semi-average by using the values of x which are below the mean ($\bar{x}$) and the values of y which are below the mean ($\bar{y}$). Use these two values as the co-ordinates of a point labelled C.

e   Draw a straight regression line on the graph guided by the position of A, B and C. These three points may not form a perfect straight line but the points can be used to obtain a more accurate placement for the trend line.

7.3 Cartographic methods

1 ISOPLETH (OR ISOLINE) MAPS

Isopleth maps are drawn to identify patterns from point data. They are based on the same principle as contour maps; the isolines are drawn to connect points of equal value and the distribution is illustrated by the pattern and density of the lines on the map. Most data can only be measured or collected at specific points in the landscape; the isoline map allows you to generalise from this point data. This technique is commonly applied to altitude, rainfall (isohyet), temperature (isotherm) and pressure (isobar) in physical geography but can be used effectively for any point data such as pedestrian counts around the CBD or for travel times (isochrones) for shoppers or commuters.

The construction of an isoline map is as follows:

a   Plot the values at each location on your base map. If your data is a value which represents an area such as the ward of an urban area then place the point in the centre of the area to which it relates.

b   Chose intervals for your isolines which are suitable for the range of data and the map area. Decide upon the number of isolines which you need to cover the range of data; this is similar to the selection of class intervals and class boundaries described on page 96 under histograms.

c   Draw in the isolines by joining points with the same value as the designated isoline. You will find that you rarely have enough points of exactly the same value so you will need to interpolate and estimate the position of the required value between points which have values on either side of the given line. It is advisable to work in pencil initially; the process is to some extent a matter of trial and error, you are likely to make adjustments as you develop the pattern.

d   A labelled isoline must pass through all the points of that stated value and must separate points of higher and lower values. Thus a line of value 15 must join all points of that value and no others; in trying to place a line of value 15 between points of value 18 and 14 then the required line would pass between the two points and would be closer to the 14 point than the 18 point.

e   Draw in your finalised lines and label them to indicate their value.

f   If you wish you can shade in the different class sizes to show the pattern more clearly; this would be done in the same way as for a choropleth map.

2 CHOROPLETH MAPS

These are basically density shading maps where the amount of colour, or the density of the shading, represents changes in the data being presented. The data is usually in the form of a ratio such as population density per square kilometre or some measure which is independent of the area of the map sub-divisions. In this way the effect of the area of the individual sub-regions is eliminated. The data must be grouped into classes, with class intervals and stated boundaries. You should refer to page 98 on histograms for some guidance on this process.

Choropleth maps are easy to construct and are visually very effective: they enable the reader to see general patterns in an areal distribution. However, they do have some limitations or drawbacks. They assume that the whole of the area has the same density and therefore show no variation within an area; this could be misleading if you use large sub-regions or a large class interval. They also suggest that abrupt changes occur at the boundaries between neighbouring regions which are placed in different classes when in reality the changes may be very gradual and bear little relation to the arbitrary boundaries on the map. Providing that these limitations are considered and allowed for in your interpretation and comments on the pattern observed, then the choropleth map can be used effectively in the description of spatial patterns.

The construction of a choropleth map is as follows:

a **Decide on the number of classes needed to display the full range of your data.**

b **Select suitable class boundaries and the range for each class (the class interval)**

c **Devise a key for the density of shading appropriate to each class; the basic principle is that the highest density should be represented by the darkest shading and that progressively lighter shading should be used for lower densities of data.**

d **Shade in each area of the map completely with the appropriate form of shading as indicated on your key.**

It is tempting to use a variety of different colours but this approach should be undertaken with great care. What appears to be a very effective form of shading in the key may not translate quite so effectively on your base map. In general softer colours such as yellow or green should be used for lower densities with areas of higher density being represented by the stronger, more striking, colours of blue, red and purple. In practice you need artistic skill to produce an effective gradation which exactly matches the key. Even if you use computer aided design methods for your project you will need to give some consideration to this aspect!

It is perfectly acceptable, and equally effective, to employ a shading technique which uses a line density form of shading in one colour. In practice it is much easier to achieve the desired result by varying the density of lines within each class. Do not leave areas 'blank' unless this is meant to represent regions for which there is 'no data'.

Make sure that lines of shading which are meant to be parallel are drawn accurately; you could place a sheet of arithmetic graph paper under your map, put this on a map table (tracing table) and use the graph paper grid as a guide in drawing the lines.

Projects in Practice

David has collected data relating to social segregation in a large urban area. The information, taken from the census, is provided for each ward of the urban area. In order to compare the distribution for two of his indicators he decides to produce a choropleth map based upon the individual wards of the city. In total there are 13 wards; using the formula to establish the maximum number of classes; i.e. 5 × (log of items), he calculates that he should not use more than five classes.

The two chosen indicators are unemployment (as a percentage of the economically active population) and overcrowding (defined as a household having more than one person per room). He has data to compare a number of time periods. The highest values are those for 1981; therefore, to enable some comparison through time his classes need to fit this data and can then be used for the other time periods in order to standardise his approach.

Unemployment for 1981 ranges from 9.1 per cent to 37.2 per cent. He uses 5 classes; 0–9.9, 10–14.9, 15–19.9, 20–24.9, and over 25. He devises a shading scheme which progresses from the highest density of line shading for ▶

Continued

'over 25' and the lowest for '0–9.9', and uses the same method for each of the 'unemployment' maps.

The same technique is applied to data on overcrowding although as the figures only range from 0.8 per cent to 12.5 per cent he adopts a system with four classes; 0–3.9, 4–7.9, 8–11.9, and over 12.

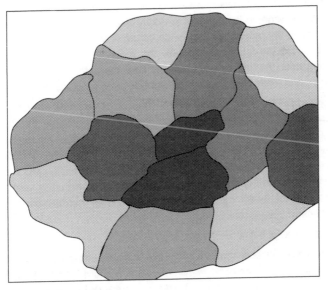

Key

Unemployment % 1981

 Over 25   20–24.9   15–19.9   10–14.9   0–9.9

Figure 7.14 A choropleth map showing social segregation in a large urban area, using unemployment figures

3 FLOW DIAGRAMS AND DESIRE LINES

These two forms of presentation are similar in that they both represent the volume of movement from place to place. The data could be in various forms:

● **traffic movements along particular routes into a city, or bus/train services**

● **inter-regional migration of population or the movement of shoppers into a major retailing centre**

● **the movement of goods or commodities e.g. the movement of oil exports on a world scale.**

In both methods the width of the line is proportional to the quantity of movement.

The difference between them is that in a flow diagram the line representing the quantity of the movement follows the actual route taken (as with bus services where the lines would follow the road network) whereas on a desire line map the line is drawn directly from the point of origin to the destination and takes no account of the specific route.

Given this difference the method of construction is identical; the choice between the two methods depends upon the particular detail of the data collected and the point which you are trying to illustrate. A flow line requires accurate data for each part of a network, for example the number of buses on each route of the network. A desire line only needs information about the origin and destination of the movement, although for some studies such as shopping catchments this may be difficult to obtain other than by means of a comprehensive questionnaire survey.

The construction of a flow/desire line is as follows:

a **Draw a suitable base map; for a flow line this will need to be sufficiently detailed to show the background network which forms the basis of the data presentation. This could be a route network on which you can mark the location of each data collection point. For a desire line, a map of the study area may be adequate.**

b **Decide upon a scale for the width of the lines. This will depend upon the full range of the values which have to be displayed. Select a scale which is easy to plot, for example 1 mm represents 100 units. The simplest form of line width scale, and the one which presents the most accurate visual representation of relative flows, is one which uses a directly proportional system. So, if 1 mm represents 100 units, 1 cm represents 1000 units.**

The choice of scale is also influenced by the amount of space available for the plots; this will reflect the density of the base network (flow line) and the area of the map/size of sub-regions (desire line).

c   Plot the data on to your base map; draw lines of appropriate width, and with parallel sides, according to your scale. It may be advisable to work in pencil initially, high volume flows may totally obscure lesser flows and you may need to modify your scale in the light of trial and error. Some flows may be so small that they may be simply represented by a single line, or you may consider them to be insignificant and exclude them. In the latter instance you should make this clear in your key by stating the threshold for inclusion in the study.

A B C D E F G H I J K L M N O P Q R S T U V W X Y Z

Analysis and interpretation of data

Statistics can be used to good effect in a project to enable more objective analysis of data and to support conclusions which may have been suggested by more subjective methods of interpretation. However, they should not be included merely as a cosmetic exercise to illustrate that you know how to work out such values; any technique which you decide to employ should be used for a particular purpose. In this way it will add to your study and form an integral part of the interpretation of data and will help you in evaluating the significance of your results.

It is essential that you:

1 select an appropriate method of statistical analysis

2 apply it correctly and work out any values accurately

3 understand how to interpret and comment upon the result.

You can use any technique which you are confident in applying. Many statistical methods are contained within computer software packages; these could be used to speed up the mathematical calculations but you still need to be aware of the relevance and the interpretation of any values calculated.

Remember, statistics only summarise data, patterns or trends; they do not necessarily prove that one variable causes a change in another. In other words, geographical interpretation is needed to make sense of statistical results.

8.1 Selecting an appropriate technique for data analysis

Many textbooks provide detailed methodology for a wide range of statistical measures which could be used in a project.

The following table indicates some of the main ways in which statistics can be used together with an appropriate method which could be applied to analyse data.

Table 8.1 Techniques for data analysis

Reason for using statistics	Method
To compare and summarise data	Measures of Central Tendency: mean median mode
To describe dispersion or variability of data	Range: upper and lower quartiles inter-quartile range standard deviation
To assess the degree of correlation between two variables	Spearman Rank Correlation Coefficient (R_s)
To assess the degree of geographical concentration of a particular activity	Location Quotients
To measure point patterns in a distribution; to assess the degree to which the pattern is random, regular or clustered	Nearest Neighbour Statistic (R_n)
To determine whether the difference between observed data (collected from primary or secondary sources) and expected data (as suggested in a hypothesis) is statistically significant	Chi Square Test

1 Measures of central tendency

a **Arithmetic mean.** The arithmetic mean ($\bar{x}$) is calculated by adding together the individual values and dividing the total by the number of values in the data set:

$$\bar{x} = \frac{\sum x}{n}$$

Although the arithmetic mean takes into account each individual value and gives a general 'average', it can be misleading if there are a small number of very high or very low values which may distort the mean.

b **Median value.** A better impression of the 'average' value may be given by the median value.

The median is the middle value in the data set when placed in rank order, i.e. there are an equal number of values above and below the median.

If the number of values in the data set is odd:

median value $= \dfrac{n + 1}{2}$th position in the rank order

Thus, if the total number of values in the data set is 21, the median would be

$\dfrac{21 + 1}{2}$th value, i.e. the 11th value in the rank.

If the number of values in the data set is even, the median is taken as the mean of the two central values.

Unlike the arithmetic mean which is an average of all the values (which may not be an actual value in the set) the median takes less account of each actual value but produces a value which will be the mid point of the overall range.

c **Mode.** The mode is the most frequently occurring value in the distribution (see page 96, histograms) and can only be identified if the individual values are known.

Each of these measures gives a different result, although the extent to which they differ depends upon the nature of the distribution; it is mathematically possible for them all to give the same value; if the data followed a 'normal distribution' then the values would be identical. With 'skewed distributions', there may be large differences between the three measures.

They each display weaknesses when used on their own. If the main purpose is to compare two sets of data then the mean may be the most useful because it does take into account every value in the data set; it therefore has a much more sound mathematical basis.

None of the measures outlined give a full picture of the data distribution, even if all three are used together. Two sets of data may have identical values for mean, median and mode but may have very different patterns of distribution about the mean.

2 Measures of dispersion or variability

a   **The range.** The range (from the maximum value to the minimum value) is a simple measure which gives some indication of the spread of the data, but again it gives little indication of the distribution of individual values.

A more useful measure of dispersion is given by the size of the inter-quartile range, the difference between the upper quartile and the lower quartile.

Upper quartile: $\quad$ UQ $=\dfrac{n+1}{4}$ th position in the rank order
$\qquad\qquad\qquad\qquad\qquad$ (rank 1 being the highest value)

Lower quartile: $\quad$ LQ $=\dfrac{3(n+1)}{4}$ th position in the rank order.

Inter-quartile range: IQR $=$ UQ $-$ LQ.

The IQR indicates the spread of the middle 50 per cent of the values in the distribution and therefore gives a measure of the degree of dispersion (see box and whisker plots on page 94).

b   **Standard deviation.** This indicates the degree of clustering of each individual value about the mean. It is calculated by first measuring the deviation (or difference) of each value from the mean; these deviations are then squared (which eliminates negative values), and then added together. This total is divided by the number of values in the data set to give the variance of the data.

Standard deviation is found by taking the square root of the variance:

Standard deviation $= \sqrt{\dfrac{S(x - \bar{x})^2}{n}}$

In a normal distribution:

68% of the values lie within ±1 standard deviation (1 SD) of the mean

95% of the values lie within ±2 standard deviations of the mean

99% of the values lie within ±3 standard deviations of the mean

Thus, if the mean rainfall was 1500 mm and the standard deviation was 150 mm:

A B C D E F G H I J K L M N O P Q R S T U V W X Y Z

68% of the data would lie between 1500 ±150 mm; i.e. between 1350 and 1650 mm

95% of the data would lie between 1500 ±(2 × 150) mm; i.e. between 1200 and 1800 mm

The standard deviation enables comparison of the distribution of values in a data set and is therefore of much greater use than the simple measures of central tendency.

A low standard deviation indicates that the data is clustered around the mean value, whereas a high value indicates that the data is widely spread with significantly higher and lower figures than the mean.

c   The coefficient of variation (*V*). This assesses the standard deviation as a percentage of the mean value.

$$V = \frac{\text{standard deviation}}{\text{mean}} \times 100$$

This is also a useful index for the comparison of two, or more, distributions because it considers the standard deviation in relation to the mean value about which it has been measured. A higher coefficient of variation would indicate that the particular distribution has a wider dispersion than a distribution with a lower coefficient.

Projects in Practice

As part of an investigation on the effect of different factors upon yields in agriculture, Roger compared two areas with very similar average rainfall figures. Despite the similarity in rainfall totals the two areas had different levels of crop yield which could be caused by a wide range of human and physical factors. In order to assess the variability and reliability of the rainfall as an input to the agricultural system he collected data for precipitation totals for a number of years for a station in each area. He then used a range of summarising statistics to compare the two areas. ▶

Continued

Roger's comparison of two climatic stations was laid out as follows:

Annual precipitation totals (in mm)

Year	Station A	Station B	Rank for B
1	1665	877	14
2	699	1082	9
3	550	1203	3
4	1188	963	12
5	1040	1241	2
6	886	1194	4
7	1091	1072	10
8	578	900	13
9	762	1146	6
10	701	1094	8
11	798	1098	7
12	1040	1318	1
13	911	791	15
14	2356	1035	11
15	1681	1151	5

Calculation of standard deviation for station B

Year	Precipitation	$x - \bar{x}$	$(x - \bar{x})^2$
1	877	200.67	40268.45
2	1082	4.33	18.75
3	1203	125.33	15707.61
4	963	-114.67	13149.21
5	1241	163.33	26676.69
6	1194	116.33	13532.67
7	1072	-5.67	32.15
8	900	-177.67	31566.63
9	1146	68.33	4668.99

Continued

10	1094	16.33	266.69
11	1098	20.33	413.31
12	1318	240.33	57758.51
13	791	−286.67	82179.69
14	1035	−329.34	108464.84
15	1151	73.33	5377.29
$\sum x$	16165	$\sum(x - \bar{x})^2$	400081.48
$\bar{x}$	1077.67	$\dfrac{\sum(x - \bar{x})^2}{n}$	26672.1
		SD =	163.32

Roger summarised his statistics for stations A and B as follows:

$$\text{Median} = \frac{n+1}{2}\text{th position in rank order}$$

$$= \frac{15+1}{2} = 8\text{th} = 1094$$

$$\text{Upper quartile} = \frac{n+1}{4}\text{th position}$$

$$= \frac{15+1}{4} = 4\text{th position} = 1194$$

$$\text{Lower quartile} = \frac{3(n+1)}{4}\text{th position} = 12\text{th position} = 963$$

$$\text{Inter-quartile range} = UQ - LQ = 1194 - 963 = 231$$

Statistic	Station A	Station B
Mean	1063.07	1077.67
Median	911	1094
Inter-quartile range	487	231
Standard deviation	476.85	163.32

Continued

Although the two stations have mean values which are similar, the spread of the data about the mean is very different; these differences could help to explain why the yields are different. The data for B is much more bunched around the mean; the middle 50 per cent of the values only cover a range of 231 mm.

The standard deviation for B (163.32 mm) is only 15% of the mean value compared with 45% for A.

For B, 68% of the values lie between the mean (1077.67 mm) ±1 standard deviation (163.32)

i.e. between (1077.67 − 163.32) and (1077.67 + 163.32)
          914.35   and   1240.99

The difference in the variability and reliability of rainfall between the two areas could explain the differences in yield; certainly it indicates that the average values may be misleading and that more detailed analysis involving correlation between yields and a number of variables may be worthwhile.

3 The Spearman Rank correlation coefficient

The Spearman coefficient is used to assess the extent to which two sets of data (variables) are correlated. It is an objective measure which provides a numerical value to summarise the relationship between two sets of data. The statistical significance of the result can be determined to assess the likelihood of it occurring by chance.

However, it must be remembered that even if there is a statistically significant relationship between two sets of data, it does not prove a causal link, i.e. it does not prove that a change in one variable has been responsible for a change in the other.

It is a useful technique for testing hypotheses; it can suggest relationships which are worthy of further investigation, or it can eliminate some variables from the study.

The test can use any ordinal data in the form of raw values, percentages or indices which can be ranked.

The formula for calculating the Spearman Rank correlation coefficient (R_s) is:

$$R_s = 1 - \frac{6\Sigma d^2}{n^3 - n}$$

where d is the difference in ranking of the two sets of paired data and n is the number of sets of paired data, i.e. the number of values for each variable.

The test effectively measures the extent to which the rank of data for one variable is different to the rank of data for the second variable and then assesses whether this difference is statistically significant.

The method of calculation is as follows:

a Rank the data for variable A from 1 to n (highest value ranked 1 to lowest value at rank n).

b Rank the data for variable B from 1 to n.

NB If two, or more, values are the same these are known as tied ranks. To allocate a rank order to such values calculate the 'average' rank which they would occupy. For example, if there were three values which should each be placed at rank 4, then add together the ranks 4, 5 and 6 and divide by 3 ($4 + 5 + 6 = 15$ divided by $3 = 5$). These would all be allocated rank 5 and the next value in order would be ranked 7 as you have already used the value which would have been ranked 6.

c Calculate the difference in rank for each pair of data.

d Square each difference.

e Add the squared differences and multiply by 6.

f Calculate $n^3 - n$; by substituting the number of values for each variable as n.

g Divide $6 \times Sd^2$ by the value for $n^3 - n$ and take the result away from 1.

The answer should be between +1.0 (perfect positive relationship) and –1.0 (perfect negative relationship).

NB The most common error made by students in calculating the R_s value is in step g; they forget to take the value calculated away from 1.

How to interpret the result:

In commenting upon the Spearman coefficient there are three aspects to consider.

a **What is the direction of the relationship?**

If the calculation produces a positive value then the relationship is positive, or direct. i.e. as one variable increases then so does the other. This would be indicated by a scattergraph with a trend line with a positive gradient. If the coefficient is negative then the relationship is negative, or inverse. A scattergraph of the two variables would show a trend line with a negative gradient.

b **Is the result statistically significant, or could the relationship observed have occurred by chance?**

In order to assess the statistical significance or the reliability of the result critical values for R_s must be consulted. Statistical Tables contain a list of critical values for Spearman.

The following table indicates the critical values at two significance levels for different values of n.

Table 8.2 Critical values for *n*

n	0.05 significance level	0.01 significance level
10	± .564	± .746
12	.506	.712
14	.456	.645
16	.425	.601
18	.399	.564
20	.377	.534
22	.359	.508
24	.343	.485
26	.329	.465
28	.317	.448
30	.306	.432

When interpreting the result for statistical significance the critical value applies whether the result is positive or negative. Ignore the positive or negative sign; simply assess the result you have calculated against the critical values given for your value of *n*.

As an illustration, if you have calculated an R_s value of 0.55 and you had 20 sets of paired data, for example 20 sample sites along the course of a river, then compare your result with the stated critical values for $n = 20$.

If your result is equal to, or greater than, the stated critical value then it is statistically significant.

0.55 is greater than the critical value at both the 0.05 and the 0.01 level. Significance is measured from 0 to 1.0; therefore 0.05 significance (which may be stated as the 95% confidence level) means that the result could have occurred by chance 5 in 100 times.

Rejection level is 0.05 significance; any result which does not satisfy this level has to be rejected; it could have occurred by chance more than 5 times in 100; this represents an unacceptable level of chance; there is likely to be some other explanation for the observed relationship.

The 0.01 significance level (99 per cent confidence level) means that the result could have occurred by chance 1 in 100 times. This means that the given result is very unlikely to have occurred by chance and represents a statistically very significant result.

Obviously, if your result satisfies the 0.01 level, then it must satisfy the 0.05 level. In this case there is no need to state that it satisfies 0.05.

You simply need to state the most significant level that your result satisfies.

c   Does the observed relationship have geographical significance?

It is important that you interpret the relationship in its geographical context; is this what you would expect? How do the two variables relate? Try to develop your reasons for the observed relationship.

You may have started the investigation by establishing a hypothesis (or null hypothesis); consider your result and relate this to your initial research question.

Points to watch for in using Spearman are as follows:

a   You must have at least ten sets of paired data, or sample points: Spearman is unreliable if n is less than ten.

b   The more sets of data used the more reliable the result; but the calculation becomes more complicated, particularly in the ranking process, and there is more chance of error once you use more than 30 sets of paired data.

c   If you have a sample which is large enough to make the calculation process unwieldy then use one of the computer statistical packages which are available.

d   If the data set contains too many tied ranks then this undermines the statistical reliability of the coefficient; there is little that you can do about the nature of the data collected but you should be aware of any limitations in the application of a chosen method of analysis.

e   Remember, the coefficient must be between +1 and –1 in value.

Without support a correlation coefficient proves very little; it is possible to identify a correlation between two sets of data without the data being connected in any meaningful way. The Spearman test merely indicates the degree of correlation between two sets of data and allows some assessment of the likelihood that the relationship occurred by chance; it does not prove that a change in one variable causes a change in the other. It does not prove a causal relationship.

It is very important that you choose your variables sensibly; do not attempt to suggest that there is a direct causal link between variables which only have some spurious connection. It may be possible to prove a statistical relationship between birth rate and car ownership per thousand population, but any observed connection is not direct, it may be caused by a third factor (such as level of GDP) which is not being investigated.

Projects in Practice

Amanda has undertaken a study to investigate the changes in a shingle storm beach in relation to distance along the coast. The beach is aligned from SW to NE and the dominant approach of the waves is from the SW. The main ridge appears to increase in height with distance from the SW and the shingle appears to decrease in size and become more rounded towards the NE. Although these general trends are easily observed on a scattergraph, there are some anomalies and Amanda decides to use the Spearman test to assess the statistical significance of the relationship.

Distance (metres from SW end of beach)	Height (m)	Mean shingle diameter (cm)	Roundness index
60	5.5	8.4	1.72
180	7.5	8.5	1.77
300	8.0	7.7	1.63
420	11.5	8.1	1.48
540	11.0	6.1	1.58
660	7.5	5.8	1.63
780	10.0	6.2	1.60
900	10.5	7.2	1.41
1020	10.0	7.5	1.36
1120	11.0	6.2	1.33
1220	14.0	6.5	1.35
1340	12.5	5.8	1.23
1460	18.0	4.8	1.18
1580	13.5	5.0	1.17
1700	15.0	5.8	1.25

The calculation of R_s between distance and the mean shingle size was completed as follows:

Table for calculation of R_s

Distance	Rank	Mean shingle size	Rank	Difference in rank (d)	d^2
60	15	8.4	2	13	169
180	14	8.5	1	13	169
300	13	7.7	4	9	81
420	12	8.1	3	9	81
540	11	6.1	10	1	1
660	10	5.8	12	2	4
780	9	6.2	8.5	0.5	0.25
900	8	7.2	6	2	4
1020	7	7.5	5	2	4
1120	6	6.2	8.5	2.5	6.25
1220	5	6.5	7	2	4
1340	4	5.8	12	8	64
1460	3	4.8	15	12	144
1580	2	5	14	12	144
1700	1	5.8	12	11	121
					996.5

$$R_s = 1 - \frac{6 \times 996.5}{15^3 - 15}$$

$$= 1 - \frac{5979}{3360}$$

$$= 1 - 1.78$$

$$= -0.78$$

The coefficient indicates that there is a negative relationship between the two sets of data; as the distance increases the mean shingle diameter decreases.

This result is significant at the 0.01 significance level; the R_s value is greater than the critical value when $n = 15$ (0.63). Such a result is very unlikely to have occurred by chance; less than 1 in 100. This is a statistically reliable result which supports Amanda's subjective perception of the longitudinal changes with distance from the SW end of the beach.

4 Location quotients (LQ)

A location quotient is a measure which indicates the degree of geographical concentration of a distribution; it produces a single numerical value which compares the concentration of an activity in a sub-region with the concentration in the whole region. It is based upon the Lorenz curve (see page 78); whereas the Lorenz curve displays the observed distribution in relation to a theoretical line of 'even distribution', the LQ presents the comparison with an even distribution as a numerical value.

The LQ can be used to measure the geographical concentration of activity, whether this is a spatial pattern or a distribution between various categories, such as the population in each social class or income group.

For example, in analysing the distribution of employment in a particular industry in the UK the formula for the calculation of the location quotient is

$$LQ = \frac{X^1/Y^1}{X/Y}$$

where X^1 is the number employed in the given industry in the region

Y^1 is the total number of all manufacturing employees in that region

X is the number employed in the given industry in the UK

Y is the total number of all manufacturing employees in the UK

The LQ effectively compares the proportion of workers in a region in a particular activity with the proportion of workers in that industry nationally.

If the data is already in percentage form then the formula is simplified to:

$$LQ = \frac{\text{\% of workers in the given activity in the region}}{\text{\% of workers in the activity nationally}}$$

How to interpret LQ values:

a The key benchmark is LQ = 1.0. If a sub-region has its 'fair share' of a distribution then the LQ would equal 1.0. A 'fair share' means that the region has a share of the distribution in the same proportion as the national distribution; this 'share' would reflect the size of the working population in relation to the size of the working population in the UK. If 10 per cent of the manufacturing workers in the UK were employed in a particular industry, then, if a region had its fair share of that activity, 10 per cent of its manufacturing population would be employed in that industry.

b If LQ > 1.0 then the region has more than its 'fair share'; this represents a geographical concentration of the activity, the activity is over-represented in that region; there is not an even distribution nationally, some regions have a greater proportion of the activity than others.

c If LQ < 1.0 then the region has less than its 'fair share'; the activity is under-represented in that sub-region.

Location quotients are useful in that they reduce four sets of data to a single, objective, numerical value and they enable comparison to be made between regions.

However, it does not indicate the number of workers in a region; it shows the proportion or concentration of workers in a given activity in relation to the national pattern.

You need to exercise care when using LQs to analyse spatial patterns; the calculated LQ is a mean value derived for a sub-region within which considerable variations are likely to occur. Such local differences may not be revealed by the data available for large areas.

The interpretation of LQs from different time periods needs careful thought; as four different variables are used in the calculation of the LQ then a change in any one of them would cause a change in the resulting value. When LQs are compared over time then all four variables are likely to have altered. A sub-region may have less employment in a given industry but the LQ may have risen because the decline of the industry in the region has been slower than the decline nationally. The relative change in each variable is important; this aspect of the LQ can cause problems in interpretation for less numerate students.

The important thing to remember is that the LQ compares the proportion in an activity in the region with the national situation at each point in time.

Projects in Practice

Gareth used data on the social class of visitors to museums and galleries as the basis of a Lorenz curve (see page 78).

Using the same data, he calculated a location quotient to measure the degree of concentration of heritage visitors in the different social classes. This was calculated by the formula:

$$LQ = \frac{\% \text{ Museum visitors}}{\% \text{ of UK adult population}} \quad \text{(for each social class)}$$

Class	% Museum/ Gallery visitors	% UK Adult population	LQ
A/B	33	19	1.74
C1	27	23	1.18
C2	24	28	0.86
D	11	17	0.65
E	5	13	0.38

This confirms the unevenness of the distribution shown by the Lorenz curve; social classes A/B are over-represented (LQ = 1.74); although they only make up

Continued

19 per cent of UK population they provide 33 per cent of visitors to museums and galleries. In contrast, social class E which makes up 13 per cent of the population only provides 5 per cent of visitors (LQ = 0.38). This could generate some debate on issues related to spending on heritage; many forms of heritage receive support from the government; all 'classes' contribute towards the upkeep of elements of our heritage through taxation but some social classes are over-represented as recipients of 'heritage investment' particularly where entrance fees have to be paid.

The application of the LQ in this way quantifies the 'unevenness' more clearly than the graph.

5 The Nearest Neighbour statistic (R_n)

This statistic, or index, analyses the distribution of individual points in a pattern. It can be applied to the distribution of any items which can be plotted as point locations. It could be used, for example, to analyse the distribution of village settlements; the location of different types of retail outlet in a CBD; the distribution of individual vegetation species.

The statistic is based upon the measurement of distance between each point in the pattern and its nearest neighbouring point. If the pattern was clustered then these distances would be relatively small; if the pattern was regular then the distances would be larger. Obviously, the spacing between points could be influenced by the size of the area being studied. In order to allow comparison between different point patterns and to overcome the differences which might be caused by variations in the size of the study area, the density of points in the area of the pattern is taken into account.

$$\text{The density of points in a pattern } = \frac{\text{number of points } (n)}{\text{area } (A)}$$

In order to calculate the Rn statistic, the measured mean distance between nearest neighbours in the pattern is expressed as a ratio of the expected mean distance between each point and its nearest neighbour. The expected mean distance is a theoretical value based upon the density of points in the area and it can be shown mathematically as:

$$\text{expected mean distance} = \frac{1}{2\sqrt{(n/A)}}$$

The Nearest Neighbour index is calculated as

$$\frac{\text{measured mean distance}}{\text{expected mean distance}}$$

And the formula is:

$$R_n = 2\bar{d}\sqrt{\frac{n}{A}}$$

where $\bar{d}$ is the mean distance between each point and its nearest neighbour

   n is the number of points

   A is the area of study

- **The index can be any value between 0 and 2.15**

- **0 represents a pattern which is perfectly clustered, i.e. there is no distance between nearest neighbours, all the points are at the same location and thus d is 0**

- **2.15 represents a pattern displaying perfect regularity; all the points would lie at the vertices of equilateral triangles forming a perfect 'nesting' pattern. Each of the distances between nearest neighbours would be identical**

- **Perfect clustering ($R_n = 0$) is not likely to be encountered since if all the points were at the same place there would be no pattern to investigate**

- **Similarly, in the real world, it is unlikely that any pattern would be perfectly regular. 0 and 2.15 form the end points in a continuum of values**

- **If the points in the distribution had occurred purely as a random pattern then the R_n value would be 1.0**

The Nearest Neighbour statistic measures the extent to which the observed pattern deviates from a random pattern and allows some assessment of the degree to which the pattern shows a significant element of clustering or regularity.

The calculation of R_n values is as follows:

a Identify the points in the pattern which is to be analysed.

For example, the pattern of villages could be shown on a tracing overlay using a common element, such as the location of the main church, to identify the individual points in the pattern. In a CBD study, the individual retail outlets could be located on a base map of the area which you have defined as the CBD; these points could then be used for the Nearest Neighbour analysis.

b Define the area of study; measure the area in the same units as those used to measure the distance between nearest neighbours. If the distance between shops is measured in metres then the area should be measured in square metres.

In some studies, such as an analysis of the pattern of shops in the CBD, the area is 'fixed' in that you have delimited the area in some way; there is a clear boundary established.

When studying settlement patterns, unless the settlements are contained in some way such as on an island, the area is less clear. You have to decide where the boundary of the study area is located. Under these circumstances it is usual to establish a 'buffer zone' around your study area; this could be a margin of two grid squares on the map beyond the established area.

When you measure the distance between each point and its nearest neighbour, if the nearest neighbour is a point in the buffer zone then you would measure the distance to that point, but that point would not be included in the study as a point from which you would measure to its nearest neighbour.

You only measure TO points in the buffer zone NOT FROM them.

c Measure the distance from each point to its nearest neighbour and record your values in a table, for example:

Point	Distance
A	
B	
C	
D	
E	
etc.	
Σd	
$\overline{d}$	

d Substitute the values for $\overline{d}$, n, and A into the formula.

Remember, your answer must lie between 0 and 2.15; if this is not the case then check your working; the most common error is that the distances and the area are in different units.

6 Chi square (χ^2)

This test is used to match categorised or classified values, referred to as the 'observed data', against a theoretical set of values, referred to as the 'expected data', in order to determine if any difference between the two is statistically significant.

The observed data is the data which has been collected; this could be based upon primary sources such as questionnaire surveys or from secondary sources such as map or documentary analysis.

The expected data is, in effect, the theoretical distribution of the data; what would be expected if a random distribution of the points or responses had occurred. It is the distribution which might be expected according to some assumed hypothesis.

Normally, before the test is applied it is necessary to formulate a null hypothesis (H_0). This is an important aspect of the test; it ensures that it is operated and interpreted in a consistent way in relation to the acceptance or rejection of the null hypothesis.

A null hypothesis would be that there is no significant difference between the observed and the expected data distribution.

For example, in a survey of shoppers who use an out of town shopping centre you could assume that there is no significant difference between the responses given by males and those given by females. This would form the basis of the null hypothesis and the actual (or observed) answers could be tested against this theoretical pattern.

Chi-square is relatively easy to apply but the data must be in the correct form and the problem should be one that is suitable for this method.

The operation of the test is fairly simple when there is only one set of observations with which to deal but it becomes much more complex mathematically when the test involves two sets of data distributions.

The operation of the test and the calculation of chi-square is as follows:

a **The easiest way to illustrate the operation of the test and the calculation of chi-square is to apply it to an observed distribution. The orientation of corries in North Wales (see page 82, Polar graphs) and in the Cairngorm region of Scotland is shown in the table.**

These represent the observed values.

Orientation (bearing in degrees from due North)	Number of corries
0–89	53
90–179	24
180–269	11
270–359	16
Total	104

The aim is to see if there is any significance in the variation of corrie orientation. If orientation had no effect on the distribution of corries then it would be expected that there would be an equal number of corrie orientations in each 90 degree sector.

The null hypothesis would state: 'orientation has no effect on the distribution of corries'.

b **The expected values would be the total (104) divided by the number of sectors (4) = 26.**

The formula for $c^2 = \sum \dfrac{(O - E)^2}{E}$

Where S means the sum of ...

O means the observed values

E means the expected values

The calculation is set out in the table below:

Orientation	0–89	90–179	180–269	270–359	Total
O	53	24	11	16	104
E	26	26	26	26	104
O – E	27	–2	–15	–10	
$(O - E)^2$	729	4	225	100	
$\dfrac{(O - E)^2}{E}$	28.04	0.15	8.65	3.85	
$\chi^2 =$	28.04	+ 0.15	+ 8.65	+ 3.85	= 40.64

In order to interpret a chi-square value the number of degrees of freedom must be determined. This is done using the formula (n–1), where n is the number of observations, in this case the number of cells which contain observed data, i.e. $4 - 1 = 3$. Statistical Tables give the distribution of chi-squared values.

With three degrees of freedom, a value of 7.82 is needed to satisfy the 0.05 probability level (the 95 per cent confidence level) which is regarded as being statistically significant. The value of 40.64 clearly satisfies this level (it actually satisfies the 0.001 level or 99.9 per cent confidence level)

This indicates that the difference between the observed situation and the expected situation is only likely to occur by chance less than 1 in 1000 times. It is very unlikely that the observed pattern has occurred by chance if there is no significant difference and the null hypothesis can be rejected; there is statistical support for a difference between the two patterns.

Points to watch when applying chi-square:

a Check that the observed and expected values are large enough to ensure that the test is valid. This is a difficult point because opinion varies as to what is the minimum number which should occur in a cell. This ranges from 5 to 20, but most sources suggest that there should be at least ten in each cell.

b Degrees of freedom is the maximum number of observations in a set of data which can vary before the value of the remainder is automatically established. In the above example there were 104 corries. Any value could be placed in three cells (three degrees of freedom) but the value in the remaining cell (16) was determined by the total in the other three cells:

$\{104 - (53 + 24 + 11)\}$

c Although a large x^2 value indicates a large difference between the observed and expected data the null hypothesis (H_0) can only be rejected after consulting statistical tables.

d A probability greater than 0.1 is not significant enough to reject (H_0).

e A probability between 0.1 and 0.05 is marginal.

f A probability of 0.05 (95 per cent confidence level) is statistically significant; the H_0 can be rejected with 95 per cent confidence, the result would have occurred by chance 5 times in 100. Any level of probability greater than this simply allows the null hypothesis to be rejected with even greater confidence.

Projects in Practice

Ian has used place-name endings to identify and locate Anglo-Saxon settlements in an area of varied relief. He uses an OS map and a geology map to divide the study area into five zones and he plots the distribution of settlements within these zones. The five relief areas, and the percentage of the area covered by each type of land is as follows:

- clay lowland 20%
- scarp foot bench 5%
- chalk scarp 4%
- chalk plateau 60%
- alluvium (in a river valley crossing the chalk plateau) 11%

His observed data is taken from the OS map.

In this application of chi-square Ian has to make an allowance for the fact that the relief zones do not cover an equal amount of the land area. In order to assess the extent to which physical background influenced the distribution of early settlement, Ian establishes a null hypothesis which states that 'physical background (relief and geology) has no effect upon the distribution of villages'.

He also finds that there are not enough villages actually on the scarp to make the test valid; rather than abandon the test, he decides to amalgamate the data for the chalk scarp with the 'scarp foot bench villages' as the settlements which are on the scarp are perched just above the bench.

If physical background has no effect upon distribution then the villages would be distributed randomly according to the area covered by each type of physical zone.

In total, Ian studied 120 villages. He calculated the expected values for each cell as follows. His expected values were:

clay: 20% of land area, therefore $120 \times 20\% = 24$

chalk bench: (+ chalk scarp) $120 \times 9\% = 10.8$

chalk plateau: $120 \times 60\% = 72$

alluvium: $120 \times 11\% = 13.2$

Ian used these expected values and his observed values to test his null hypothesis.

In both of the previous illustrations the mathematics has been relatively straightforward. Under certain circumstances the application of the chi-square test can become more complex; if you find the mathematics difficult, then ask for help or use a different form of analysis.

The test is more complicated if there are two sets of observations or if there is only one degree of freedom.

Both of these situations affect the calculation of the expected values. The observed data is not affected; this is whatever has been collected as observed data.

Projects in Practice

Anne has conducted a questionnaire survey of shoppers to determine the impact of an out-of-town shopping centre. Some of her questions have specifically looked at the responses of males and females in order to see if there is any difference in the two groups with regard to the changes in shopping. The responses to a question, as either a negative or positive answer, are shown for males and females in the following table:

	Yes		No		Total	
Males	41	a	49	c	90	(x)
Females	68	b	40	d	108	(y)
Totals	109	(X)	89	(Y)	198	(Z)

In order to see if there was any significant difference in the response of males and females Anne established a null hypothesis which stated that 'the sex of the respondent had no effect upon the answer'.

Assuming this to be the case, the expected values were calculated. Anne had to allow for the fact that there were a different number of males and females and also a different number of Yes and No answers:

- the individual cells in the table have been labelled a,b,c,d to show the observed data
- the ROW totals have been labelled (x) and (y)
- the COLUMN totals have been labelled (X) and (Y) for reference.

Continued

Because of the difference in the number of males/females and the replies Yes/No, the calculation of the expected values for each cell (a, b, c, d) becomes more complicated:

- mathematically, you have to calculate the probability of values occurring in cell a, and then do the same for b, c, d
- the probability of values falling in the Yes column is 109/198, i.e. the total number of Yes replies in relation to the Total number of replies
- the probability of values falling in the Male row is 90/198, i.e. the total number of males in relation to the total number of replies.
- therefore, the probability of values falling in the Male and Yes cell (cell a) is $90/198 \times 109/198 = 0.25$.

If the total number of replies (198) is multiplied by this probability then the expected value for cell a is 49.5

For cell b: $108/198 \times 109/198 = 0.30 \times 198 = 59.5$

For cell c: $90/198 \times 89/198 \quad = 0.204 \times 198 = 40.5$

For cell d: $108/198 \times 89/198 \quad = 0.245 \times 198 = 48.5$

Once this principle is understood there is a shorter version of the calculation:

- the expected frequency for a given cell is given by multiplying the column total by the row total
- and then dividing by the overall total.

For cell a Column total (X) × Row total (x)

$$= \frac{109 \times 90}{198} = 49.5$$

Overall total (Z)

This is then applied to each cell in turn.

$$b = \frac{(X) \times (y)}{Z}$$

$$c = \frac{(x) \times (Y)}{Z}$$

$$d = \frac{(Y) \times (x)}{Z}$$

Continued

In this case there is also only one degree of freedom: in a matrix of this type the number of degrees of freedom is determined by $(n - 1) \times (n - 1)$ where n is the number of cells in each axis of the table i.e. $(2-1) \times (2-1) = 1 \times 1 = 1$

It is obvious that if you were to change the value in cell a, then all the other cells would change providing there was no change in the total number of responses.

When you are dealing with one degree of freedom it is necessary to adjust the observed values by 0.5 in the direction of the expected values; this is known as Yates Correction.

Anne's working table was adjusted to include the expected values and the 'corrected' observed values.

So, the Males/Yes cell is adjusted from an observed value of 41 by 0.5 in the direction of the expected value, i.e. by 0.5 towards 49.5; a value of 41.5.

	Yes	**No**	**Total**
Males	$O = 41.5$ $E = 49.5$	$O = 48.5$ $E = 40.5$	90
Females	$O = 67.5$ $E = 59.5$	$O = 40.5$ $E = 48.5$	108
Totals	109	89	198

Chi-square is the sum of $\dfrac{(O - E)^2}{E}$ for each cell in the table.

$$= \frac{(41.5 - 49.5)^2}{49.5} + \frac{(67.5 - 59.5)^2}{59.5} + \frac{(48.5 - 40.5)^2}{40.5} +$$

$$\frac{(40.5 - 48.5)^2}{48.5}$$

$$= 1.29 + 1.08 + 1.58 + 1.32$$
$$= 5.27$$

With one degree of freedom; the critical value for a probability of 0.05 is 3.84.

Therefore, the null hypothesis can be rejected; there is less than a 5 in 100 likelihood that the difference between the observed and the expected data are due to chance. There is a significant difference in the male and female responses; this could suggest that they have a different perception of the impact of the shopping centre.

The important point to remember with all of these statistical and analytical techniques is that they are only there to be used as support for your own ideas on the geographical significance and relevance of the material which you have collected as part of your study.

It is vital that you relate your results and statistical analysis to your initial hypotheses and that you evaluate your findings in the context of established theories in that area of the subject.

Your results may support the established ideas, but equally you may observe some anomalies which in themselves might suggest other research questions which could be investigated to gain greater insight or understanding of the topic.

Making some geographical sense of your results is more important than simply displaying the mathematical ability to calculate statistics.

✓ Checklist 3

Am I writing up the investigation clearly?

You should now be at the stage of writing up the investigation. These are the questions that you should now ask.

	Yes	Partly	No
1 At the beginning of the report is my aim(s) clearly stated?	☐	☐	☐
2 Have I made the link between my investigation and the wider part of the subject with which it is associated?	☐	☐	☐
3 Have I referenced all sources of ideas and information?	☐	☐	☐
4 Have the methods that I used to collect the data been clearly described and their use justified?	☐	☐	☐
5 Has the area of my investigation been clearly located on a map?	☐	☐	☐

Continued

	Yes	Partly	No

6 Have I made reference at the appropriate places in the text to my illustrative material so that they form an integral part of my report? Does the information that they convey agree with what is written in the text?

7 Are the photographs that I have included relevant to the investigation and not merely 'space fillers' to make my completed report look good? Have I correctly annotated or labelled my photographs?

8 Are my interpretations of the data valid and are my results summarised in the conclusion? Have my conclusions been linked back to my aims and have I indicated how far these aims have been achieved?

9 Have I made a thorough evaluation of the investigation

10 Have I written an executive summary? (see Glossary)

The writing process

There are three main factors to consider in order to give the report of your investigation a sound framework, clear style and attractive appearance:

* **structure**
* **language**
* **presentation.**

9.1 Structure

You need to give form and shape to your report. A basic structure helps the reader digest the report and it helps you to write and organise your material in a logical manner. A general structure for a report is given below but you must remember that there may be particular sections that have to be included depending on the examination board to which you are submitting the completed piece of work. It is therefore a very good idea to check the appropriate syllabus document before you embark upon the writing process.

Here is a typical report structure:

* **report approval/cover sheet followed by title page and contents**
* **executive summary**
* **aims and objectives**
* **statement of research questions, issues, hypothesis or problem**
* **sources of information used, methods of collection and analysis and their limitations**
* **analysis and interpretation**
* **evaluation and conclusion**
* **bibliography and appendices.**

The first thing that you should do is concentrate on writing the body of the report. This is the introduction, the findings and the conclusions. After this, deal with the other sections. The following order for writing is suggested:

1 Analysis and interpretation

This is the section in which you present your findings. When you are writing this section all your material should have been sorted, selected and arranged in note form. This section includes:

i   the results of your analysis

ii   your interpretation of those results

This section forms the basis for your conclusions. You should help the reader by ending each separate section with its own conclusion, This will show what you judge to be the major factors and gives an easy way to refer back in the reading to earlier passages within the report.

2 Methods

In this section you should discuss:

i   the sources of evidence that you have used

ii   how you have collected and analysed the evidence

iii   the limitations of the sources and methods of collection and analysis.

3 Conclusions

This includes the summary of all the major findings made at stages throughout the report. No new evidence should appear at this point. The conclusion considers the evidence presented in the main body, draws out the implications and brings it to one overall conclusion or an ordered series of final conclusions.

4 Introduction

After writing your findings and conclusions you should now know clearly what you want to introduce. The introduction is where you acquaint the reader with the purpose of the report and guide them through the structure of it.

5 Appendices

This section is set aside for supplementary evidence not essential to the main findings, but which provides useful back-up support for your main arguments.

6 Contents

All the sections of the report should be listed in sequence with page references.

7 Bibliography

This section covers the books, magazines and other sources which have been used in your research. It must include every reference mentioned in the text and be presented correctly. This means that you should give the author, initials, the title of the book (or the articles title and the magazine it appeared in), the page number and date of publication. It is also good practice to include the name of the author in the appropriate place in the text as a direct indication of some relevant background reading. These names in the text should then be referenced again in the normal way in the bibliography.

8   **Title page**
     This should include the title of your report, which indicates its central theme. Title pages should also include your name and the date of the completion of the report.

9   **Executive summary**
     This is a very important part of the report and should be the last thing that you write. The best way of going about constructing such a summary is given on page 145.

9.2 Language

Remember that first impressions count; it is unwise to put the reader off before they have even studied the report.

You are solely responsible for what you write and for the words you choose to express your thoughts. Remember that although you may have an individual style of expression this does not excuse poor English. Your style will not necessarily be immediately apparent to the reader, but poorly expressed English will. Your sentences must be grammatically correct, well punctuated and words must be spelt accurately. Poor writing often indicates muddled ideas. You do not really know what you are saying until you put it into words that another person can easily understand. Remember you are writing to communicate, not to perplex or impress. Focus on the specific purpose of the report and avoid jargon; every part of the report should relate to this purpose and such an approach will help to keep the report concise and coherent.

Accuracy is vitally important so always be precise, and ensure that you are using the correct words as clarity is essential. Do not write phrases or sentences that may have more than one meaning. To avoid this, know precisely the material that you are trying to convey.

Keep sentences short and simple. Long complex sentences slow the reader down and confuse and impede understanding. The same applies to paragraphs.

Poor spelling automatically detracts from your work and generally annoys the reader. Use a dictionary if you are uncertain or a spellchecker on a PC if you are using one. (NB Be aware of American spellings on some spellcheckers.)

In the overall assessment of your work the examination boards have to take into account your spelling and grammar, so with poor English you will undoubtedly receive a lower mark than you would have been given if attention was paid to this area. AQA, for example, in Specification B, assesses the quality of your language, by insisting that to obtain full marks at any level, the appropriate Quality of Language descriptor must be achieved by candidates as follows:

Level I   Style of writing is suitable only for simple subject matter. Clear expressioin of only simple ideas, using a limited range of specialist terms. Reasonable accuracy in the use of English.

Level II   Manner of dealing with complex subject matter is acceptable, but could be improved. Reasonable clarity and fluency of expression of ideas, using a good range of specialist terms, when appropriate. Considerable accuracy in the use of English.

Level III Style of writing is appropriate to complex subject matter. Organises relevant information and ideas clearly and coherently, using a wide range of specialist vocabulary, when appropriate. Accurate in the use of English.

9.3 Presentation

Your finished report must look good in addition to reading well. Adequate headings and numbering make it easier for the reader to comprehend what you are saying. This stage of report writing requires the sane level of care that went into composing the text.

The presentation of statistics is often more informative and eye-catching if it is done visually, for example by using graphs, pie charts or histograms.

Layout is very important, this is the relationship between print and space on the page. This applies whether your report is handwritten or word processed. A crowded page with dense blocks of print or writing and little space looks very unattractive and is off-putting. When organising your text always ensure that you have provided:

- **adequate margins**
- **either double or 1.5 spaced lines**

- headings that stand out clearly from the page
- paragraphs that are relatively short
- plenty of space between paragraphs and sections.

9.4 Writing an executive summary

An executive summary is defined as 'the main points of a report'. The purpose is to provide the briefest possible statement of the subject matter of a longer document and must:

- remain faithful to the original report
- cover all the essential points
- be fully comprehensible when read independently of the full document.

Your executive summary must not be a list of extracts, highlights, or notes on the original.

When writing the executive summary it is important to remember that:

- it must introduce the subject of the full report, its objectives, methods, findings and/or recommendations
- it must help the reader to determine whether the report is of any interest.

The writer can use this summary as a rigorous check on the success of the full report, i.e. is the report clear and concise and does it meet its aims.

The method of constructing an executive summary is as follows:

a Read the whole document.

b Isolate and summarise its central theme.

c Read each section and identify and summarise the main findings or points.

d Combine b and c into a set of major points because your aim is to convey the overall impression of the full document in as brief and clear a way as possible.

e Read through your summary to check that it gives a fair impression of the original while ensuring that it will make sense to the reader as a separate document from the full report.

9.5 The word limit

All boards state the length of report that they wish to see and are prepared to impose some sort of penalty on overlong submissions. Edexcel, for example, in Specification A states that candidates must present a report that should 'be no more than 2500 words in length' and that 'enquiries outside this range cannot achieve full marks as detailed in the mark scheme'. The mark scheme in this case only allows a report to be credited up to 6 marks under the heading of 'Investigation Design and Planning' for which there is a maximum of 12 marks. On the B Specification from Edexcel at A2 level, the coursework report should be 'about 1500 words' but 'for reports in excess of 20% overrun of the word limit you will not be awarded any marks for that part of the study that exceeds 1800 words' and you will be expected to complete an accurate verification of the number of words that you have used. In the report you do not have to count tables, diagrams and statistical workings within the total words. Limits are not designed to make life difficult for you as the evidence is that studies which keep within the limits are better written and balanced than those that stray beyond them.

9.6 Pre-submission editing

It is important not only to read the draft through from start to finish before submission but also to edit and refine the material.

As you read, mark pages that will need attention later but do not stop to deal with them at that moment. You need to get a feel of the overall structure and impact of the report first so your initial read through must be continuous. Put yourself in the reader's shoes and be highly critical of what you have written.

Proofreading is vitally important. Regardless of the time and effort put into writing the report, the required result will not be achieved without sufficient care being devoted to proofreading. A poorly typed or handwritten report, full of errors and inconsistencies in layout, has a damaging effect regardless of the quality of the content.

Here is a summary of points that you need to cover and to ask yourself at this stage in the completion of your report:

* **the report needs to be checked in great detail for grammar and spelling errors**

* **ask yourself whether you could have expressed yourself in a better way. If so, change the sentence or even paragraph**

- assess whether the structure of the main body of work is really the most suitable one to present your material, ideas and arguments

- is each paragraph well structured?

- are all the maps and diagrams included in the correct place and integrated into the text?

- are all the references in the text included in the bibliography?

- does the report fulfil the stated aims and assessment objectives?

- is your argument watertight and easy to follow?

- does your conclusion make your argument all the more convincing?

- does your executive summary convey the key points of the report?

- assess the layout and general appearance of the document.

You are now in the position of being able to submit your report with the confidence that it is well done and carried out to the best of your ability.

A B C D E F G H I J K L M N O P Q R S T U V W X Y Z

✓Checklist 4

Am I ready to hand in my work?

You should now be at the stage where the report on your investigation has been written and you are in a position to hand in the complete piece of work. Before you do, ask yourself these questions.

	Yes	Partly	No
1 Have I carefully read the report and ensured that everything is in the right order and that there are no errors in grammar, spelling and the typing?	☐	☐	☐
2 Have I kept to the word limit and where necessary ruthlessly edited the material?	☐	☐	☐
3 Have I completed a contents page and properly paginated the whole report?	☐	☐	☐
4 Is there a bibliography in which I have included all details of source materials used and persons consulted?	☐	☐	☐
5 Have I put any excessive information into an appendix along with copies of my recording sheets?	☐	☐	☐
6 Have I included at the front of the study a copy of my original proposal form with comments from moderator/adviser (depending on requirements of board)?	☐	☐	☐
7 Is the whole report enclosed in the type of folder which is acceptable to the board in question?	☐	☐	☐

10 Answering examination questions based on fieldwork and practical skills

We indicated earlier that all AS/A2 geography specifications require students to undertake investigative work. Most students will carry this out and eventually submit the finished piece of work for marking. The specification, to which you are working, may allow this marking to be carried out internally within your own institution or may require you to submit the work to an external examiner appointed by the awarding board (AQA, Edexcel, OCR). Some specifications, however, allow you an alternative. You will still be required to carry out a fieldwork investigation, but will not be required to submit a finished document. Instead you will be entered for an examination in which you will be asked various questions about your investigation. Other questions may test fieldwork-related or investigative skills. These are often based on data provided by the boards, in some cases as pre-released material, and candidates are asked to present, analyse and evaluate that material. In other words, these questions assess the candidate's practical abilities.

When being assessed by fieldwork-related questions, the following areas of your investigation could form the basis of the questions (some typical questions have been included to give you a flavour of what can be expected).

- **The aims of your enquiry.**
- **The size and limits of your study area. ('Describe and justify how you defined the geographical limits of your study area.')**
- **The hypotheses/research questions that formed the basis of your investigation. ('State the hypothesis or research question which you established as part of the enquiry and explain why you chose it.')**
- **The data to be collected and the methodology of the collection. ('Describe the data to be collected and the methods used to collect that data.' OR 'Analyse the limitations of the methods that you used to collect the data and the precautions that you took to ensure that the data were as accurate as possible.' OR**

'How did you ensure that the data you collected were the most appropriate to the ideas/hypotheses you were testing?')

- **The methods used to present the data.**

- **The methods used to analyse the data.**

- **A knowledge of your results and the extent to which they supported your hypothesis/research questions. ('Discuss the extent to which the results of your enquiry supported your hypotheses/research questions.')**

When responding in the examination to the above material, the following points should be remembered:

Aims, hypotheses, research questions: your response will depend upon the marks awarded. At its simplest, you should state the aim and the hypothesis, etc, but if it is clear that a more considered response is required (more than one mark for a hypothesis, for example) then to access these marks, more contextual detail would have to be offered.

Study area: all geographical studies have to be conducted within boundaries, which have to be both appropriate and practical. Here, you should demonstrate that you selected an appropriate study area, and justify why the chosen area was right for the methodology of your enquiry. It would be wise to explain the processes that led to the selection of the area of data collection, preferably with reference to more than one set of criteria. For example, such criteria could include the ease of access to the area in question, the need for some uniformity of sampling population (say an area of common rock type) or, conversely, a need to sample a wider variety of data sources (for example, questionnaires in a number of different settlements). High marks would certainly be awarded if you could demonstrate an understanding that scientific enquiry demands careful control of the study area – it should not be too small or too large, and should have clear and relevant parameters.

The data and its collection: both the data and the method of collection must relate to the hypotheses/research questions and be totally practicable. You should also show a clear awareness of the limitations of your data collection process, as well as describing practical problems that may have been encountered. The best answers would also include some evaluative commentary on the inherent problems of fieldwork design and experimentation. It is important that you recognise that problems of accuracy exist and that improvements could be made under ideal, but often unrealistic, circumstances. Precautions to ensure accuracy could relate to methods used to ensure accurate instrument readings, or the need for the averaging of several readings. Accuracy could be improved by better design of data collection methods, for example, sampling techniques or questionnaire design.

Data presentation: the chosen methods of data presentation must link logically to the data collected, and be appropriate for those data.

Data analysis: as above, the chosen methods of analysis must be suitable for the data collected. It is important that an effective means should be clarified by which the conclusions can be generated from the analysis. The method of analysis should follow on logically from the data collection and original hypothesis.

Results: you must show some appreciation of the geographical significance of the results of the enquiry, and of how they add to the understanding of the environment studied. References to results have to be provided, together with statements of conclusions that have been developed. Reference to anomalous results should be made, together with some evaluation of the overall success or otherwise of the enquiry. The best candidates do not try to bluff the examiner, they give accounts of accurate results and findings, and state when and where they went awry. It is very rare for a geographical enquiry to go completely to plan!

It is also possible, on some specifications, for the examiners to set a broader question that covers much of the above. An example of such a question could be:

‘● **Geographical research commonly follows a series of steps like those outlined below:**

● **hypothesis or idea**

● **experimental design (sampling, measurement)**

● **data collection**

● **results and analysis (statistical test, etc.)**

● **reconsideration of experimental design**

● **theory and explanation.**

Discuss the extent to which you followed similar steps in your enquiry. Where you deviated from them, explain why.’

In the question there are six steps outlined, but your answer should concentrate on three basic areas. First, you should link the steps given in the question to your enquiry in a clear way. There should be a discussion of your enquiry with exemplification throughout. In other words, the context of your enquiry must be straightforward to see and relate to the steps given. Second, the steps given in the question must be followed as a complete research project. In particular, to achieve the highest mark, the fifth step, which recognises the need for review and evaluation, must be clearly and demonstrably understood. Finally, valid explanations for any deviations from the steps to the enquiry must be given, together with an explanation that a more full and detailed application of the steps may have led to an overall improvement.

Guide to investigations: AQA, EDEXCEL, OCR

This section sets out what you need to know to get a high grade on your coursework depending upon the unitary awarding body to which you are submitting the work. It is written on the assumption that you have read the previous chapters of the book.

11.1 AQA Specification A (submission at A2 level)

The nature of the fieldwork investigation

You are required in Assessment Unit 6 to produce a fieldwork investigation at a local, small scale within which primary data collection must take place. Such investigative work in the field should develop skills associated with planning, collection of primary data (and secondary, if required), presentation, interpretation and evaluation. Some classroom-based background study will be necessary to support the investigative work. Before you make your final choice there are several very important points of which you should be aware:

- **there are no restrictions on the type of topic studied, other than it should include primary data collection and should be based on a small area of study**

- **any geographical argument, assertion or problem may be investigated**

- **the completed investigation has a recommended word limit of 4000 words**

- **the work should be based on a minimum of two days spent in the field**

- **the selection of a title for the investigation should be one that is manageable and that can be fully developed within the word limit**

- **when devising the aims, you should select a focussed hypothesis, issue or aim that has both a theoretical and locational context as a large number of related hypotheses are** unlikely to permit you to complete the investigation within the word limit.

* methods of data collection should be set up so that they are relevant to the aims and they should be manageable by you in the time available

* your collected data should permit the use of appropriate cartographic, graphical and statistical skills in order that a full interpretation can be made

* your interpretation should include reference to the aims

* your conclusions should include a summary of the results, the relevance of these to the aims and an evaluation of the overall investigation.

Procedure

In your A2 year, you should seek advice on your choice of topic from your teachers to ensure that you are able to show what you understand and what you can do. If there is some uncertainty then it is possible for your teachers to seek advice from the AQA's Fieldwork Advisory Service. The AQA accepts that group work may provide a useful basis for the undertaking and teaching of fieldwork, but this should not necessarily lead to the production of similar investigations within the same centre. Your contribution and individuality should be demonstrated in the finished investigation.

You are allowed to consult your supervising teacher during the course of your investigation but he/she has to record what took place and inform the Authority accordingly. You should also be aware that you have to acknowledge in your final version all assistance and sources of information that have been used.

As part of your work it is likely that you will search and utilise information assembled by others. There is, however, an important distinction between plagiarism and the acquisition of information by research. The distinction lies in the use made by candidates of the information they obtain and the acknowledgements of the sources used. You must use quotation marks when a direct quotation is made from written source material; when other material is used, you must make sure that it is properly referenced. If you are not careful in both respects you may be accused of deliberate deception. You will be required to sign, as at GCSE, a declaration of authenticity.

The investigation may be handwritten or prepared using information technology, although you will not gain additional credit solely for IT use. Where IT is used for data presentation, the quality, range, relevance and understanding of the techniques will be the main assessment criteria.

Your investigation will be externally assessed by an Authority appointed moderator/examiner. The AQA will notify you of the date by which you have to submit the completed investigation through your teacher. Your work should be compiled into a folder of A4 size with individual sheets numbered, secured together and identified with your centre and candidate number. You should not insert your material into plastic pockets or submit the final work in a ring binder.

Approaches to fieldwork investigation

During your investigative work, it will be expected that you show an ability to do the following:

* **display an understanding of the purpose of the investigation and its relevant background, both in a spatial and conceptual sense**

* **show an awareness of the suitability of the data collected and the methods used**

* **be able to evaluate the methodology**

* **be aware of alternative methodology**

* **use the collected information in a straightforward way, presenting it in a different or more easily understood form such as graphs or maps**

* **be familiar with alternative methods of data processing/presentation**

* **describe, analyse and interpret data in relation to the aim**

* **draw conclusions relating to the specific enquiry and understand their validity and limitations**

* **use and understand your own experience of fieldwork and enquiry**

* **plan, construct and carry out sequences of enquiry**

* **demonstrate an awareness of safety issues and risk assessment in geographical fieldwork.**

Fieldwork opportunities from the Specification

Sections of the Specification that could be considered for fieldwork (with some suggestions) include:

Water on the land	Changes to the main river parameters downstream.
	Comparative studies of river channel cross-sections.
	Evaluation of a river management scheme.
Climatic hazards and change	Microclimate studies e.g. urban heat islands.
	Acidification of rainfall.
Energy and life	Studies of soil characteristics, catenas, etc.
	Vegetation succession studies using transects.
	Footpath erosion combined with soil compaction studies.
Population dynamics	Local small-scale migratory movements.
Settlement processes/patterns	Identification of the characteristics of a commuter village.
	Impact of gentrification in a small area.
	Studies of urban land use, patterns
	Studies of the range and threshold of services/centres.
Economic activity	Study of some form of local primary activity in light of its environmental impact.
	Changing fortunes of a local industrial estate.
Coasts/processes/problems	Changes of beaches/dunes/etc. over time.
	Impact of human activity on coastal systems.
Managing cities	Decay, deprivation and the evaluation of the success of contrasting strategies.

Dynamic nature of the CBD.

Urban ecology.

Recreation and tourism

Study of a honeypot and the environmental impact upon it.

Ecological capacity of a local tourist attraction.

Analysis of changing patterns in local tourism.

How will your investigation be assessed?

Your completed investigation will be externally assessed by AQA. Your school/college is required to collect your work, authenticate it and then submit it to an examiner by a date, which you will have previously been given.

The examiner will assess your work under the following headings:

* **aims (10 per cent of the total marks awarded)**

* **methodology (25 per cent)**

* **skills used (25 per cent)**

* **interpretation (25 per cent)**

* **communication skills (10 per cent)**

* **conclusions (5 per cent).**

Aims

To achieve the maximum credit (10 marks) you must show a very well focused aim to your investigation. The context of your work must be very good in both a theoretical and locational sense and these elements must be linked clearly and effectively.

Methodology

Maximum marks (25) will be awarded for a clear and detailed summary of methods of data collection. Your collection of primary and secondary data must be seen to be rigorous and well linked to the aims of the investigation. Sampling must be fully understood and explained, and any piloting well applied. A good awareness must be shown of the limitations of the methods of data collection. If the data have been collected in a group of candidates, then individual flair must be apparent.

A B C D E F G H I J K L M N O P Q R S T U V W X Y Z

Skills

To gain the maximum 25 marks you must have made very good use of the relevant cartographic, graphical and statistical skills and techniques. Significance must be fully understood and explained. If you use CAD, your approach must be seen to be competent and relevant to your investigation.

Interpretation

To obtain the highest credit (25 marks) your interpretation must make strong references to the aims of your investigation and its theoretical and vocational contexts. Skills and techniques will be well integrated to aid the interpretation. Anomalies will be well explained.

Communication skills

In order to gain the maximum 5 marks you will need to show detailed and sophisticated communication skills with a cogent writing style and excellent use of geographical language.

Conclusions

The maximum mark in this section is 10 and to achieve this you will need to write a very good conclusion including references to your results and to the original aims/theory. Any evaluation must be seen as relevant and self-critical and you may well offer constructive proposals for further development of the study.

11.2 AQA Specification B (submission at A2 level)

The nature of the fieldwork investigation

You are required in Module 7 to submit a written investigation based upon an enquiry (a question, a hypothesis, a problem) into a geographical issue linked to the specification content at either AS or A2. This module assesses you on the enquiry process, from its inception to the writing up of the report. Your work will be internally assessed by the teachers within your centre and a sample of work from that centre will be submitted to an external moderator appointed by AQA, who will review the work to determine whether any adjustment is necessary to bring the centre's assessment into line with wider standards. To be able to complete this investigation successfully, you should be able to:

- understand how to identify a clear enquiry question/ hypothesis/problem and how to develop aims from this

- identify, select and collect (using a range of techniques) appropriate quantitative and qualitative evidence from primary sources (including the field) and secondary sources

- present data in an appropriate manner

- analyse data

- evaluate results

- write an investigation report on an enquiry.

Advice on setting up the investigation

Before you make the final choice of subject for your investigation, you should be aware of the following:

- There should be sufficient source material available to study the topic chosen for the investigation, and the investigation should be capable of being completed in the time available.

- The investigation should consist of approximately 3500–4000 words (exclusive of bibliography, maps, diagrams, and other illustrative material).

- The collection of evidence can be carried out in a group, but some of the design of the enquiry and the whole of the execution of the investigation must be your own, unaided work.

- Assistance and all sources of information that have been used must be acknowledged at the end of the investigation. There is an important distinction between plagiarism and the acquisition of information by research. This distinction lies in the use made by you of the information you have obtained and the acknowledgement of your sources. You must use quotation marks when a direct quotation is made from written source material and when other material is used, you must make sure that it is properly referenced. As with GCSE projects, you will be required to sign a declaration of authenticity.

- In group activity you must indicate the part you played in the collection of evidence.

- You may need to adjust the emphasis or direction of the enquiry once it is in progress.

- You should consider a range of sources which could include some of the following: textbooks, biographies, diaries, monographs, TV programmes, statistical data, newspapers, maps, novels, government papers, museum and field sites.

Content and structure

Your investigation should:

- involve the use of appropriate geographical skills

- involve the use of evidence from primary sources, including fieldwork

- be well structured and organised; the writing style in which the investigation is submitted (e.g., a report style in which every section is numbered) is less important than its organisation.

The investigation should be organised in the following ways:

- It should include a statement of the aims and objectives, and of the question, issue, hypothesis or problem forming the subject of the investigation.

- The main body of the investigation must include details of the sources of information used, the methods used to collect and analyse evidence and their respective limitations, followed by an analysis and interpretation of the subject of the investigation.

- The investigation should include evaluation (the degree to which the original aims and objectives were realised) and conclusions (including their limitations, implications and questions generated).

- You should present a bibliography which must include details of all source materials used and persons consulted, with an indication of the help or information received from them in the preparation of the investigation (this material does not contribute to the word recommendation).

- You may make use of a computer or wordprocessor to access, manipulate and present information. Credit for such information and the use to which it is put, however, will be confined to the extent to which you have chosen, interpreted and analysed material appropriate to the subject under study and against the assessment criteria stated in the specification. A statement of which computer software has been used must be made in the bibliography.

Presenting your completed investigation

There are several important points to bear in mind before you present the completed work:

* each page must be A4 in size and have a margin to the right of the text of at least two centimetres

* each page and each illustration should be numbered consecutively

* typewritten or wordprocessed reports should use double-spacing between the lines

* a completed Candidate Report Form should preface the investigation. This should be followed by a title page and a page listing the contents, and the latter should include lists of section headings, diagrams and graphs, etc.

* the report should be presented in a lightweight folder with contents loosely fixed; plastic wallets for individual pages should not be used.

Assessment

Your completed investigation will be assessed under the following headings:

* knowledge (10 per cent of the marks)

* critical understanding (30 per cent)

* knowledge and critical understanding (20 per cent)

* skills and techniques (40 per cent).

To obtain full marks at any level, the appropriate Quality of Language descriptor must be achieved. At the highest level this includes: the style of writing being appropriate to complex subject matter, that the information and ideas are organised clearly and coherently, that use is made of a wide range of specialist vocabulary when appropriate and accurate use of English.

Knowledge

For the highest level of marks (7–10) you will have needed to demonstrate relevant geographical knowledge at a high level. Throughout your investigation you should have used the appropriate geographical terminology and have accurately located the area of study and provided some relevant geographical characteristics. You must also provide details of the geographical ideas, concepts, principles or theories, on which this study is based.

Critical understanding

At the highest level of marks (21–30) you should show by explanation that you understand the geographical knowledge at this level. You must also provide some explanation that the sources of evidence, concepts and principles/theories available for this enquiry have limitations and/or an explanation of their potential. Where appropriate, you should have shown by explanation, an understanding of the effects of attitudes, values, approaches, decision-making processes and/or the effect of some greater involvement of human activity.

Knowledge and critical understanding

For the highest mark level (9–20), your interpretation and analysis of the data must be detailed with the correct use, where appropriate, of basic statistical techniques. You should also carry out a detailed synthesis from relevant geographical facts and ideas. From the data, you should develop conclusions with some development of their limitations and give a clear and detailed evaluation of the success, or otherwise of the investigation.

Skills and techniques

In order to qualify for the highest level of marks (16–40), you must use and demonstrate most skills and methods appropriate to your enquiry and have used most of the potential sources. You must also have correctly used some high level collection and sampling skills along with higher skills in organisation and presentation. Your investigation must include a detailed evaluation of the appropriateness and limitations of the skills used and the evidence gathered, including a clear awareness of the relevance of scale (temporal and/or spatial) in the selection of location and the collection of data.

11.3 Edexcel Specification A (submission at AS level)

Nature of the personal enquiry

The personal enquiry can be based on any part of Specification A (AS or A2). You are required to undertake a small scale investigation which must enquire into a topic, question or issue relevant to the subject criteria for geography. It must include the collection of primary data based on fieldwork and your direct practical experience. Data from secondary sources should be used where and when appropriate. You will be expected to demonstrate an ability to collect, select and interpret information geographically and represent it by a range of graphic and cartographic means. Through writing up your enquiry you will have an opportunity to show an in-depth understanding of the chosen topic. It should enable you to develop your geographical understanding of one or more aspects of the specification and to explore interrelationships between different strands of geography.

Designing and planning the personal enquiry

You are required to seek approval for your coursework plan before beginning work upon it. Your teacher should provide you with a coursework proposal form which your centre will send to Edexcel for approval on a date which will be given to you by your teachers. Edexcel will either approve your proposal, approve it with some suggested modifications or require you to resubmit the proposal. In the last case, Edexcel will indicate to you where there are flaws in the proposal. The proposal form will contribute to the assessment of marking criterion under Investigation design and planning.

When you design and plan the enquiry it is important that you take into account the following:

- **the enquiry should be investigative, in that a question or problem should be identified so that the work has a clear focus**

- **the work should be analytical; it should attempt to seek explanations for observed data in depth rather than breadth**

- **the enquiry should be manageable, in that it needs to be conducted on an appropriate scale, where there is accessible and available data for you to collect**

- **it has to be individual, as you are required to measure data, use a variety of field methods, techniques and analysis, and formulate conclusions (you may collect data, particularly at the earlier stages, as part of a group but the enquiry must have a clear individual outcome)**

- **it must meet all the assessment criteria and the standards of communication required at AS level.**

Completing the enquiry

Your report should:

- be written on A4 size paper secured together in a lightweight folder with your name, number and centre name and number on the front; your original coursework proposal form should be included

- consist of text supported by relevant maps, diagrams, tables, and other illustrations, all at A4 size (or folded to fit in the folder)

- be no more than 2500 words long (you cannot achieve full marks if you exceed this limit)

- be legibly handwritten, typed or wordprocessed

- contain well presented illustrations demonstrating your ability to create maps and diagrams (you can include computer-generated graphics)

- indicate, with quotation marks, where you are directly using written source material produced by others (there is an important distinction between plagiarism and the acquisition of information by research); all such material must be properly referenced and, as at GCSE, you will be required to sign a declaration of authenticity

- contain a bibliography

- contain an appendix only where absolutely necessary; such material should generally be incorporated within the body of the text.

How will your enquiry be assessed?

Your completed enquiry will be assessed by the teachers in your centre using the criteria set out below. A sample of the work of your centre, however, will be sent to a board-appointed moderator who will review the work to determine whether an adjustment is necessary to bring the centre's assessments into line with wider standards.

Your work will be assessed under the following headings:

- **Investigation design and planning (20 per cent).**
- **Data collection (15 per cent).**

- **Data presentation (15 per cent).**
- **Analysis and interpretation (30 per cent).**
- **Conclusions and evaluation (20 per cent).**

You will also be assessed on the quality of your written communication. Under Investigation design and planning you should be able to 'select and use a form and style of writing appropriate to the purpose of your investigation and to complex subject matter' and you must also 'organise your relevant information clearly and coherently'. Under Analysis and interpretation, and Conclusions and evaluation you must use 'specialist vocabulary' and show an ability to 'ensure that text is legible, and spelling, grammar and punctuation are accurate so the meaning is clear'.

Investigation design and planning

To achieve the highest level of marks here (10–12) your proposal form must be well organised with a clear and realistic plan of how the fieldwork is to be carried out. In your enquiry there must be a full statement of the aims and you must also show an ability to identify geographical questions/issues of your own. Aims should be clearly linked to relevant theoretical background and the location should be set in its geographical context. Your enquiry must be within the 2500 word limit.

Data collection

For the highest marks in this section (7–9) you must be seen to have followed a systematic research programme and have collected sufficient data to meet the aims of the enquiry, making a range of accurate observations and measurements. The methods of collecting data, including sampling, should be justified and significant factors affecting them taken into account.

Data presentation

The highest level of marks (7–9) is awarded when you can show that you are able to select an appropriate and varied range of presentation techniques which may include cartographic, graphics diagrammatic and statistical techniques. They must be executed precisely, accurately and with understanding, and be suitably labelled and annotated. Such material must be well integrated into the text and you must also give a well organised and logical presentation of the data collected.

Analysis and interpretation

To achieve highest marks (14–18) you must produce an effective, coherent and independent analysis and interpretation which draws upon all the information collected and presented and is directly related to the stated aims. Significant interrelationships and patterns should be identified and developed. Statistical analysis and significance testing should be accurately used when appropriate. If you have collected data as part of a group, there must be clear evidence of individual analysis on your part. The range of appropriate geographical terminology that you use must be good and there must be very few, if any, errors in grammar, punctuation or spelling.

Conclusions and evaluations

At the highest level (10–12 marks) you must have drawn on all of the information gathered to provide answers and reach conclusions, which are expressed clearly and succinctly. Your conclusions should relate directly to the aims of the enquiry and be derived from the data collected, and they must be justified. You must also show an awareness that such conclusions may be partial, tentative or incomplete. Any evaluation must be of the enquiry as a whole and 'where relevant' there should be comment on the opportunities to extend the study. A good range of appropriate geographical terminology should be used and, as in the proceeding section, you must make few errors, if any, in grammar, punctuation and spelling.

11.4 Edexcel Specification B (submission at both AS and A2 levels)

Submission at AS level

Nature of the investigation

You are required in Unit 3 to submit a fieldwork investigation of a site or small scale area in one of the environments studied in Units 1 or 2 of the specification. Your proposal must be approved by Edexcel. You are required to develop a research action plan, which follows a route to enquiry on an issue or question, chosen by you and arising out of the fieldwork. You are also allowed to access secondary sources.

The main points therefore, that you must consider before starting your investigation are as follows:

- The focus of your study has to be linked to one or more of the environments studied in AS Units 1 and 2 (i.e. river, coastal rural or urban) and be an environmental investigation that links environmental geography with physical and/or human processes.

- The recommended length of your report is 2500 words, excluding maps, data, diagrams, tables and the appendix. If you go over 3000 words you will be penalised. There is no minimum length, but reports should follow the full route to enquiry as failure to do so will result in heavy penalties.

- Evidence must be shown of primary investigation via relevant fieldwork (one to two days) supported by the appropriate use of secondary sources.

- Your proposal must be approved by Edexcel.

- You are allowed to carry out group work when collecting primary material but you must individually develop a question, issue or hypothesis which you wish to investigate, and formulate an action plan to develop a route to enquiry.

- the focus of your investigation must be small in scale.

The Environmental investigation has been designed by Edexcel to do the following:

- to provide you with an opportunity to undertake personal investigative fieldwork and gain an understanding of appropriate methodologies for primary and secondary research, and of the potential limitations of geographical data

- to develop knowledge, understanding and skills appropriate to the investigation and give you an opportunity to develop sustained enquiry

- to form an integral part of the teaching syllabus as linking Units I and 2 enabling you to accumulate knowledge, understanding and skills to focus on an issue of your choice

- to enable you to have maximum opportunities for the development and acquisition of key skills; your action plan can be used as evidence as it enables you to record significant progress on your investigation and to have this progress verified internally by your centre.

A B C D E F G H I J K L M N O P Q R S T U V W X Y Z

Framework for carrying out the Environmental investigation

1   Approval of your focus by Edexcel (approval within one month of submission).

2   Preparation for fieldwork and site visit:
    * linkage with theory
    * data collection techniques discussed and decided
    * make a fieldwork plan
    * consider advice on safety.

3   Fieldwork:
    * primary data collection (you must provide evidence of your involvement)
    * data processing
    * collation of data by groups or individuals
    * issues identified.

4   Develop your individual research action plan as you write up the report:
    * identification of issues for focus
    * development of route for enquiry (one or two key questions)
    * setting scene – summarise key geographical aspects of selected question, issue or problem and place it in its geographical context
    * data collection – assessment of methodology
    * data representation – use of a range of appropriate techniques
    * data analysis – use of a variety of techniques (graphical cartographical statistical and written)
    * evaluation and conclusion – may include management plan
    * presentation and summary of findings of the investigation in a clearly documented and well structured report which should include a summary analysis of wider significance.

5   Submit your report, with a fully documented action plan, for marking and moderation on the date given to you by your teacher. The final report should:
    * be word processed unless you have sought an exemption from Edexcel (you should retain a copy of your report on disk)
    * be placed in a lightweight A4 folder
    * have all maps, diagrams, etc. folded down to A4 size
    * only use plastic wallets when the material is fragile or complex
    * have an appendix containing evidence of individual data collection – this should contain only limited evidence, such as a sample questionnaire response

- contain details of any video or audio evidence – do not send the original
- acknowledge the use of data contributed by other candidates and material that has been directly obtained from secondary sources; you must use quotation marks when a direct quotation is made from written source material and when other material is used, you must make sure that it is properly referenced, (there is an important distinction between plagiarism and the acquisition of information by research and the former may result in strict penalties being applied); as with GCSE projects, you will be required to sign a declaration of authenticity
- have the completed individual research action plan form (form GB2) included at the front, together with the proposal form (form GB1).

Assessment of your investigation

Your completed report will be marked by the teachers within your centre. A sample of the work from your centre will be sent to an Edexcel appointed moderator, who will review the work to determine whether any adjustment is required to bring the centre's assessment in line with wider standards. Your work will be assessed under the following headings:

- **Definition and the context of a question, issue or problem (15 per cent)**
- **Methodology of data selection, collection and recording (15 per cent).**
- **Data representation (10 per cent).**
- **Data analysis and explanation (20 per cent).**
- **Conclusion and evaluation (25 per cent).**
- **Quality of written communication (15 per cent).**

Definition of a question, issue or problem

To obtain the highest marks here (13–15) you must have a very clear definition with detailed location and the geographical context must be very well explained.

Methodology of data selection, collection and recording

The highest marks in this section (13–15) are given to material, which shows a relevant, evidenced, comprehensive data collection programme. The techniques which are used must be appropriate with sound methodology. The presentation is very well organised and initiative is shown.

Data representation

To obtain 9–10 marks, which is the highest level, you must have a wide range of well presented, appropriate data representation techniques.

Data analysis and explanation

The highest mark level (17–20) is awarded to material in which there is precise explanation and a detailed analysis of all results using a wide range of techniques. You must show a clear understanding of overall trends and make an appropriate use of models and theories to inform data analysis.

Conclusions and evaluation

To obtain the highest level of marks (21–25) you must write a very thorough conclusion with a detailed evaluation of your investigation, and a high quality summary of its wider significance. You must also show a sound understanding of how to develop an enquiry.

Quality of written communication

At the highest level (13–15 marks) you must have shown a very well organised, logical route to enquiry, with excellent standards of spelling, punctuation and grammar. Your use of terminology must also be precise.

Submission at A2 level

You are required in Unit 5 to produce a coursework report of about 1500 words on one of the options selected from 'Challenges for Human Environments' which are Development and disparity, Feeding the world's people, Health and Welfare, and The Geography of Sport and Leisure.

Approval and completion

The title of your coursework is selected, devised and researched by you, but it must be submitted to Edexcel for approval on the date given by your teacher (Edexcel will reply within three weeks). The title should focus on one or more of the generalisations shown in the option selected within the specification document.

You should carry out appropriate research, which may be of a primary and/or secondary nature, and write your report in the form of a formal essay. The report should contain a full bibliography and research should be properly referenced. As with AS level, you are required to wordprocess your report unless you have an exemption from Edexcel (you must retain a copy of your report on disk). On submission your report should:

* **have the original, assessed work stapled together with the front sheet (Form GB5)**

* **have all diagrams, maps and graphs folded down to A4 size**

* **not contain plastic wallets except for fragile or complex material**

* **include the completed Form GB5 at the front**

* **include an authentication statement, signed by you, that it is your own work**

* **have some signed verification from you on the number of words contained in the work; if your report runs 20 percent over the limit (excluding tables, maps, diagrams and statistical working) you will, as a penalty, not be awarded any marks for that part of your study which exceeds 1800 words.**

Assessment

Your work will be externally marked by an Edexcel examiner. He/she will assess your work under the following headings:

Introducing, defining, and describing the question, problem or issue, and identifying the data/information required to answer it (16.66% of the marks)

To be awarded the highest marks in this section (8–10) you must give clear statements of the questions, problems or issues posed. The material must also contain the nature of the data needed and why, with reference to a range of locations and/or scales. Key terms must be accurately defined.

Researching relevant sources, selecting appropriate case study material and using this knowledge in detail (25%)

To reach the highest marks (13–15) a wide range of reading and research must be evident. Your use of excellent case studies covers a wide range from a variety of locations and scales with plenty of evidence of careful and relevant selection of material. You must make appropriate use of annotated maps, figures and diagrams which accompany the written research and are incorporated into the answer. Your research sources should be evidenced.

Understanding of general concepts, case studies, attitudes and values, and the application of data and information to the question, problem or issue (25%)

Within the highest levels of marks (13–15) you will have to show an ability to organise data logically throughout and also show that you can apply research material fully to the question, problem or issue. You must show an appreciation of a range of values or perspectives relating to the issue or problem. Most or all of your data should lead directly to the question and be well sequenced and explained. Your results are highly cogent and the answer, well structured.

Drawing appropriate conclusions on the basis of evidence, and on-going evaluation (16.66%)

The highest marks here (8–10) are given when a conclusion is clearly stated and is directly related to the rest of the essay. Your conclusions or ideas drawn should generally draw on evidence given in the essay, and you will have grasped the complexity of the question, problem or issue. These conclusions should have been built up progressively through the essay.

Quality of written communication, including the communication of knowledge, ideas and conclusions in a clear and logical order, and the use of appropriate geographical vocabulary (16.66%)

To obtain the highest marks (8–10) your material should be well written with a clear sense, coherence, and style and with clear syntax. The organisation of material should be into sequenced paragraphs which lead from one to the other with clear and logical arguments developed as the essay progresses. Knowledge and ideas should be integrated and lead to conclusions. Your terminology includes appropriate use of specialist vocabulary.

11.5 OCR Specification A (submission at both AS and A2 levels)

Submission at AS level

The nature of the Personal investigative study

In Unit 2682 you are required to carry out one investigation leading to a report of not more than 1000 words. The writing of this report, and the participation in the fieldwork on which it is based, will provide you with the background for formulating investigative questions at the local level. It is also intended for you to gain first-hand experience of data collection, and the analysis and presentation of data, which will be examined in the Unit 2682 examination paper. The investigation should utilise both primary and secondary data and may focus on a topic in either physical, human or environmental geography and it should relate largely to the local scale so that you will have the opportunity of understanding at first hand the relationship between your chosen topic and the real world. Your completed report cannot be taken into the examination room, but the examination consists of one question containing stimulus material at the local field scale, which will relate to your report. The completed report itself must be submitted to OCR along with your examination script.

Completing the investigation

The five major steps that you should undertake in completing your investigation are as follows:

1   **Identify the question(s) to be asked and any further necessary formulation into approaches that can be researched.**

2   **Development of a strategy to answer the question(s).**

3   **Collection of the data that are needed, with due regard to sampling schemes, and the organisation and recording of the data into appropriate forms, together with any cartographic, diagrammatic or numerical representation.**

4   **Analysis, evaluation and interpretation of the data to answer the initial question(s).**

5   **Presentation of a summary of the final answers or conclusions with a judgment by you on their reliability and a consideration of limitations identified by you in the preceding stages of the investigation.**

In the content of your investigation you are expected to demonstrate the knowledge, understanding and skills involved in:

●   **the identification of the types of question that can be asked in geographical investigations and the ability to formulate hypotheses where appropriate**

●   **the planning of the organisation of an investigation, considering both theoretical and practical issues including risk assessment. The importance of background reading should be understood** along with the need for development of operational plans that reflect available resources

● the existence of a continuum of data types from primary to secondary and of a range of both quantitative and qualitative methods of primary data collection and recording. You should become familiar with sampling methods and the need to consider the reliability and accuracy of the data. You should also be able to represent this data in the most appropriate form by using such methods as pie charts, histograms, scattergraphs and simple line graphs, and in map form with appropriately selected scales and symbols (O.S. maps at 1:50 000 and 1:25 000 should be considered)

● the use of appropriately labelled and annotated photographs

● the representation and exploration of data by numerical methods such as simple descriptive statistics including measures of central tendency (mean, median, mode) and measures of variation (range, interquartile range, standard deviation). You should also be familiar with tests for investigating the difference between samples and others that test association, such as Spearman's rank correlation

● investigating the value of satellite images, air photographs, geographical information systems, simple statistical software and information and communication technology

● the use of data, together with appropriate text to answer the question originally asked

● the selection and use of only the appropriate data

● summarising the findings in a conclusion and commenting on the significance of findings in the light of the reliability of the data and its accuracy

● the development of a critical examination of all aspects of an investigation.

How will your investigation be assessed?

Of the 90 marks available for Unit 2682, your report carries 10. To achieve the highest level of marks (7–10) your study must be organised under the five major steps listed above under Completing the investigation, and utilise both primary and secondary data. Your work must be clearly expressed with the correct use of geographical terminology and should be almost entirely free of errors in all sections.

Submission at A2 level

The nature of the Personal investigative study

In Module 2685 you are required to submit a research assignment, in the form of a Personal investigative study of 2500 words, on a geographical topic of your choice, although it should normally be related to some area of Specification A. It could be on a topic similar to that which you chose for AS level (Unit 2682) but must be based on material not previously submitted for examination.

Completing the personal investigation

The investigation must be based on your own collection, handling and interpretation of primary data. The study should be based on either your own fieldwork, or primary data derived from documentary sources. Primary data derived from documents could include population census, local archives, newspapers and weather satellite data. Appropriate selected secondary data may be used to supplement your primary data. Where data derived from sources other than fieldwork are used, they must be acknowledged.

Your work should usually be based on a clear geographical question that allows you to conduct a complete investigation, including its design and execution, analysis, its reporting in essay form, and an evaluation of the limitations of the essay and investigation. You are expected to devote an amount of time to the acquisition of the necessary skills and techniques, and to the investigation and its reporting, that reflects the 15 percent of the assessment that it carries at A2. Your completed investigation should not exceed 2500 words and must be accompanied (as at GCSE) by a declaration that it is your own work. You are advised to consult the assessment criteria to check that there are no major omissions in the completed study.

Assessment

Your completed investigation must be submitted to OCR and they will assess it in the following ways (all five sections are marked out of a maximum mark of 18 marks):

Formulating a question or hypothesis capable of being researched and understanding the limitations imposed on geographical enquiry by the available resources, including data. Designing realistic strategies, including risk assessment.

To obtain the highest level of marks here (15–18) you are expected to have formulated a clear geographical question or hypothesis that is expressed in researchable terms and is of an appropriate scale and feasibility. Your explanation of the reasonableness of the question or hypothesis must be well expressed in terms of the geography of the chosen theme, and your arguments must be well argued, in detail, and correct. The question or hypothesis must be set in the relevant geographical theory, with appropriate reference to published material where relevant and available. Your approach-plan to be adopted in the investigation is made clear and indicates an ability to devise a detailed and carefully planned approach that is practicable in resource terms. Your proposals for time management must be sensible and should be apparent. You must have developed a detailed risk appraisal. Throughout your text you must use geographical terms correctly and it must be clear that detailed planning has taken place with the appropriate use of reconnaissance and trials.

Carrying out programmes of data collection using selected sampling techniques

To reach the highest levels of marks (15–18) you should show in your data collection a balance between primary and secondary material that is well matched to the chosen question or hypothesis and that collection should have proceeded in a logical and carefully planned manner. Your sampling framework should have been correct for the circumstances of the chosen investigation. The amount and nature of the data collected should have been matched to the question or hypothesis under investigation and some attention should have been paid to the accuracy and reliability of the data collected. If necessary you should have repeated measurements and observations and shown evidence that the data has been the subject of monitoring and consideration during the collection period.

Representing the data using the most appropriate methods

To receive 15–18 marks, the form of the data presentation you have used should be entirely appropriate to both the data collected and the means of data analysis employed. Variety of presentation will only be rewarded if appropriate. If you use maps they must conform to the accepted geographical standards in respect of scale, keys and other properties. Your graphical representations should be clear and accurate in allowing observed data to be presented at the same level of accuracy as they were observed, and appropriately labelled for later reference in the text. If you are making comparisons, scales must be chosen which facilitate this. All your illustrative material of a pictorial nature should be annotated where appropriate and clearly labelled. When using secondary sources, they must be clearly acknowledged and a clear indication provided of the extent to which data have been modified by you. You must make sensible decisions as to where to use computer-generated presentations and where to use handdrawn alternatives.

Analysing the data using appropriate methods

To reach 15–18 marks, you must use appropriate analytical methods which are closely linked to the methods of data collection and presentation. Those methods must be correctly selected and used. Interpretation should be correct and the end product linked to the aims of the investigation. No superfluous analysis should be presented, and obvious conclusions should be accepted as such without redundant testing.

Drawing conclusions and the critical evaluation of their significance and reliability

The highest mark level (15–18) is reached by showing that the conclusions relate clearly to the original question or hypothesis base of the investigation and that the circle is closed explicitly. The question posed should be answered or the hypothesis evaluated in terms that consider the degree of success achieved. Explanations are offered to explain the extent to which original aims were met. Explicit use must be made of your findings from the analytical section and points made must be illustrated using the material gathered. Your conclusions must contain cogent arguments where necessary in the evaluation of results, rather than proceeding by assertion. Your conclusions must also contain a weight of material that is appropriate to its assessment weighting.

The alternative to Module 2685

As an alternative to Module 2685, you may take Module 2686 which concerns investigative skills. As with the similar paper at AS level, you are required to submit along with this paper a 1000 word report on an enquiry relating to an issue, question or problem involving first-hand data collection in the field and its analysis, leading to conclusions. Your completed work is handed in with your examination paper. No part of the work must involve material previously submitted for assessment in the AS Unit 2682.

11.6 OCR Specification B (submission at both AS and A2 levels)

Submission at AS level

The nature of the geographical investigation

For Module 2689, you are required to carry out a geographical investigation in which you should acquire and develop a range of subject-specific skills, which should be selected for their appropriateness to the investigation itself. The full list of skills that could be considered are contained within the Specification document. As with the Module 2682 on Specification A, your completed investigation must be submitted to OCR along with your examination paper on which there will be questions referring to your completed report. When completing your investigation you should bear in mind the following:

- It must consist of not more than 1000 words.

- Not more than two A4 size figures may be included to demonstrate your presentational skills and to illustrate key aspects of the investigation.

- It must be accompanied by a brief description (not more than 100 words) summarising the work on which the report is based. This will provide the examiner with an outline of the work undertaken.

- You should show how decisions were made about the design, conduct and presentation of the investigation.

- You should show the outcomes of the investigation.

- You should make an evaluation of the investigation.

- It is possible to produce a report based on two or more short pieces of work or a single, longer piece.

- It is an opportunity to demonstrate your skills in investigation through first-hand data collection, as required by the subject criteria.

- Topics should be chosen from those within Modules 2687 and 2688 of the Specification.

- There are no formal procedures for approval or guidance, although written advice can be obtained from OCR.

- **The work in the field can be done on a group basis, but your report must be a personal commentary on the work undertaken.**

- **Within the specification document there are questions at the beginning of Modules 2687 and 2688 which could become suitable starting points for you to devise more specific questions.**

Submission at A2 level

Nature of the investigation

In Unit 2690 you are required to produce a report on a geographical investigation which should arise out of, but extend, the content of one or more of the modules in the Specification. The topic that you choose could be an extension of the work that you have already undertaken for the AS level in Module 2689. There are two alternative routes that you can choose to pursue:

a The investigation is based on data collected in the field and should be supplemented by other primary and secondary material. This could include the use of information technology where appropriate.

b The investigation is based on primary and secondary data, data collection in the field not being a requirement. Significant and appropriate use must be made of information technology and communication technology in order to gather, process, analyse and present the data.

Both routes must produce an investigation which demonstrates a range of geographical skills; a list of skills is given in the Specification document, these being additional to the skills listed for AS level. Data collection may be done in groups but the title that you choose must be distinctive and the procedures and design must be based on your own decisions. Any collaboration in the early stages of the investigation must be recorded and justified.

The completed report should not exceed 2500 words plus maps, diagrams, etc. and it will be internally marked within your centre with external moderation carried out by OCR.

Planning and completing the investigation

There are no formal procedures for approval of your investigation. When making your plans, whichever route you intend to follow, it is a good idea to include the following:

- identification of a question, or issue for investigation, leading to a title for the investigation

- identification of relevant sources of information, their definition, potential and limitations

- identification of the scale, time and place of the investigation

- identification of the geographical dimensions of the investigation and where they fit within the broader fields of study in the subject

- identification of the ways in which relevant information is to be collected and recorded, including sampling strategies if appropriate

- how the information may be tabulated, processed and represented including appropriate maps and diagrams

- proposals for the analysis of information including statistical techniques if appropriate

- a provisional framework for the writing of the report and timescale for its completion

- awareness of potential errors or bias in the information and ways in which these might be reduced

- ways in which conclusions might be reached in relation to the original question.

If it is your intention to go down route **b**, then to meet the requirements with the regard to information technology you will need to undertake all, or most, of the following activities:

- select and extract information from electronic sources such as e-mail, the Internet, direct fax or CD-ROM

- select and use appropriate software, for example spreadsheets, databases, wordprocessing desktop publishing packages and multi-tasking applications

- edit and process information, making calculations where necessary

- combine information from different sources.

How will your completed investigation be assessed?

Your report will be marked by the teacher in your Centre and standardised by all of those teaching at this level. After the marks are submitted, a number of reports will be sent to an external moderator appointed by OCR, who will review them to ensure that the standard for the award of marks in coursework is the same for each Centre. The teachers in your Centre will assess your coursework under the following headings:

- **investigation design and data collection (21 marks out of a total of 90)**
- **data recording, processing and presentation (21 marks)**
- **data analysis and evaluation of methods (21 marks)**
- **geographical understanding and application of knowledge and critical understanding to unfamiliar content (27 marks).**

Investigation design and data collection

To reach the highest level of marks here (15–21), you will need to identify a context for your investigation and be able to select and justify the best focus for study. You must identify all the relevant data needed and select a variety of sources and methods of collecting data appropriate to the focus. In the field or from information technology sources, you must systematically, and with accuracy, collect a range of relevant data and research several secondary sources, incorporating the evidence from the relevant ones. You must show initiative, imagination and problem-solving skills in developing the investigation.

Data recording, processing and presentation

The highest marks here (15–21) are awarded for you showing initiative and using a wide range of appropriate techniques to record and present data. The information that you have obtained must be processed carefully and presented in the form of tables, maps, graphs and other diagrams which are completed accurately and, where appropriate, visually, in order to communicate effectively. You must sequence the investigation in a logical and clear way and present it in good English, using geographical terminology, maps and other graphic forms of presentation with confidence. You must also be seen to use an appropriate structure and style of writing effectively, and this must include an acknowledgement of all sources.

Data analysis and evaluation of methods

For 15–21 marks you must be able to analyse and evaluate the investigation in a concise manner, and interpret and analyse the data to arrive at meaningful and supportable conclusions. Your observations must be seen to be perceptive and you must make an intelligent analysis of the results using a wide range of different techniques. You will have to show an understanding of the value of the collected evidence to the focus of the investigation as well as its limitations, and be perceptive in the evaluation of error and bias where it appears in collected information. You must evaluate the conduct of the investigation in terms of the original goals, the reliability, validity and usefulness of the findings, and how it might be extended, and justify the selection of the data collected, the methods of collection and the presentation used.

Geographical understanding and application of knowledge and critical understanding to unfamiliar content

At the highest level (21–27 marks) you will need to demonstrate an impressive understanding of the principles and methodology of geographical investigation, and understand the purposes and limitations of the data collection and processing techniques employed. You must recognise the geographical dimensions of the investigation, as well as its wider relevance, and have applied a number of concepts and theories to the interpretation of the evidence. You must also have applied an impressive geographical knowledge and understanding to your chosen context and understand that there may be alternative interpretations of the evidence. If there are any unique aspects of the investigation results, then you must have identified them and put forward some explanation, as well as doing the same for the expected or typical results. You should be able to draw concise conclusions from the information analysed, and understand how different elements of geography may be relevant to the investigation.

A-Z glossary of geography coursework

aim: the beginning of your geographical investigation where you should be making a statement as to what you are trying to do. This could be in the form of testing a hypothesis or establishing research questions.

annual abstract of statistics: a book of key economic and social statistics produced annually by the government. It is an important source of secondary data such as population figures. The abstract is produced by the Office of National Statistics (ONS) and available from HMSO Books and in most public libraries.

appendices are pages at the end of a document where you can place supporting evidence which is too bulky or detailed for the main body of text. Good use of appendices ensures that the reader is not bogged down with page after page of numerical data. On a questionnaire, for example, it is sometimes a good idea to put the responses to some of the questions in an appendix in the form of lists making sure that the reader knows where to find them.

bias: deflection from the truth as a result of an individual's self-interest or faulty research methodology. Bias can occur in two main areas of your investigation:

1 **when sampling it is possible for the sample to become unrepresentative of the population in question as a result of poor choice of method or the selection of an insufficient number**

2 **biased views obtained when interviewing.**

Bias, if identified, should always be discussed in your final presentation.

bibliography: a list of your written research sources such as books, newspapers, magazines or leaflets. They should be provided in detail, with the author name, the publisher and the date of publication. A bibliography will impress only if its contents have been referred to directly within the text of the report, therefore it is good practice to give the occasional direct quotation.

closed questions: a question to which a limited number of pre-set answers are offered, e.g. 'Do you buy your bread from Town A?' Yes/No.

correlation: measures the degree of association between two sets of data involving the comparison of one set of variables with another either by the use of a correlation coefficient such as Spearman's or the construction of a scatter graph.

evaluation: a judgement based on evidence. This may be made by drawing conclusions from the evidence presented throughout the investigation and/or as conclusions at the end.

evidence: facts, figures or the views of others which lend support to your conclusions.

executive summary: the main points of a report. The purpose of writing an executive summary is to provide the briefest possible statement of the subject matter of the report as a whole. It must therefore cover all the essential points and be fully comprehensible when read independently of the full report.

external moderation means that the coursework will be evaluated by an external examiner, not someone from your own school or college.

feasibility check: a study to see whether or not your investigation can proceed to a satisfactory conclusion.

footnotes: explanatory notes put at the bottom of the page and referred to with an asterisk or superscript number at the end of a word or sentence. An explanation of an obscure term would be an example of where you could most profitably use a footnote.

hypothesis: a theory that can be tested through research. Many investigations are based around such testing. A typical title for your investigation in the form of a hypothesis could be: 'The distribution of facilities for old people in Bolton does not match the demand for their care'.

methodology: the method(s) by which you intend to use to achieve your objectives.

moderator: a person appointed by your examination board who could be responsible for approving your proposed investigation and for judging on a sample of your centre's work that the examination standards have been met when the reports have been internally assessed by your school/college. The examination boards have different procedures and you should consult the relevant chapter in this book for more information as to your particular circumstances.

null hypothesis: a negative assertion which states that there is no relationship between two variables which are being tested, e.g. 'there is no relationship between building height and distance from the centre of the CBD'. It assumes that there is a high probability that observed differences between two sets of data are due to chance variations. If the null hypothesis can be rejected, then we can assume that any differences between the two sets are not due to chance but are the results of differences between the two populations. We could then suggest that the hypothesis was acceptable.

objectives are targets which must be achieved if the overall aim of the investigation is to be reached. Once a student sets out the objectives of an investigation then the appropriate methods for achieving them can be decided.

open questions are those that invite the respondent to answer freely and therefore give a greater choice than closed questions. On a questionnaire they are questions that do not have boxes to be ticked. Answers can be quoted directly in your report but the results can often be difficult to process.

piloting a questionnaire means testing it out on a few people before finalising the questions. In this way questions which are unclear can be identified and changed. This particularly applies to postal surveys.

plagiarism is including in your final written version material which is not your own work but has been copied directly from a book or magazine article. You can include such material but you must acknowledge the fact that you have done so and indicate the source(s) used.

primary research: collecting information yourself by the means of fieldwork. It is an essential component of all A Level investigative work. This can be done through questionnaires, interviews, river and beach measurements, traffic and pedestrian surveys, etc.

questionnaire: a document containing a series of questions designed to discover the information required to meet your research objectives. The information is normally obtained by face-to-face interviews with a sample of the population under study but it is also possible to leave your questionnaire with respondents for later collection or to use the postal service.

research questions are very similar to hypotheses except in this case the statement is put in the form of a question, e.g. 'What factors determine the distribution of old people within Bolton?'. This could, like a hypothesis, become the title of an investigation.

sample size: the number of interviews or objects to be measured in your investigation. The key is to ensure that a large enough sample is taken for meaningful conclusions to be drawn.

sampling is the collection of data from a fraction of the total population, the results of which are then said to be typical of that population as a whole. For example, predictions of voting patterns before an election are made upon the results of the sampling of a small number of the electorate. Sampling is usually the basic method of data collection in investigations at this level. There are various methods of sampling, geographers using mainly random, systematic or stratified systems.

secondary data is that which has been compiled by someone else and that you have access to as part of your investigation. Such material will have been compiled in written, statistical or mapped form. All investigations should include some references to secondary material if a high mark is to be achieved but investigations based solely on secondary data are not allowed at Advanced level.

sensitive questions are those that should be avoided when compiling a questionnaire. These are usually questions on age, salary and education. Rather than asking a person's age, divide ages into groups and ask about the type of employment rather than the remuneration. You could also ask a person at what age they ceased full-time education.

sources: clear statements of where information has been obtained, particularly secondary material. If, for example, you use material from a text book, it is essential that you identify both the book and the author.

synthesis: the drawing together of the evidence collected as part of your investigation which leads to the formulation of a conclusion.

word limit: the approximate maximum word count allowed by your examination board. As the boards have different regulations, it is essential to check as there may be some penalty imposed for exceeding these limits. Usually this does not include tables of figures, diagrams or appendices.

A-Z of coursework sources

This section gives you information regarding some of the books and magazine articles that could be consulted during the preparation and execution of your coursework. We have put these into subject areas to make it easier for you to use.

1. Text books

Title and Author	Publisher and Publication Date	ISBN
Techniques in collection, analysis and presentation		
Advanced Geography Fieldwork, Frew J.	Nelson 1993	0 17 438 4920
Fieldwork – Firsthand, Glynn P.	Crakehill Press, 1988	0 907105 20 3
Fieldwork Techniques and Projects in Geography, Lenon B. & Cleves P.	Collins, 1994	0 00 326643 5
Skills and Techniques for Geography A-Level, Nagle G. with Witherick M.	Stanley Thornes, 1998	0 7487 3188 1
Physical (general)		
The Physical Environment 16–19 Geography, Digby B. (Ed)	Heinneman, 1995	0 435 35227
Geomorphology and Hydrology, Small K.J.	Longman, 1989	0 582 35589 3
Integrated Physical Geography, Taylor J.A.	Longman, 1994	0 582 35583 4
Coasts		
Coasts, Hanson J.	Cambridge Univ. Press, 1988	0 54 313777 5

Hydrology

River Basin Management, Doherty A. & McDonald M.	Hodder & Stoughton, 1992	0 340 55394 4

Environment, biogeography and soils

Soils and Environment, Ellis S. & Mellor A.	Routledge, 1997	0 415 06888 6
Environmental Pollution, Foster I	Oxford Univ. Press, 1991	0 19 913368 9
Biogeography, Tivy J.	Oliver & Boyd, 1993	0 582 08035 5
Techniques in Fieldwork in Ecology, Williams G.	Collins, 1991	0 00 322246 2

Human (general)

Understanding Human Geography, Bradford M & Kent A.	Oxford Univ. Press, 1994	0 19 913310 7
New Patterns Process and Change in Human Geography, Carr M.	Nelson, 1997	0 17 438681 8
The Human Environment 16–19 Geography, Digby B. (Ed)	Heinneman, 1996	0 435 35226 1

Urban geography

Urban and Rural Settlements, Carter H.	Longman, 1990	0 582 35585 0
Urban Social Geography, Knox P.L.	Longman, 1995	0 582 22937 5
Land and the City, Kivell P.	Routledge, 1993	0 415 0872 1

Rural (settlement and agriculture)

Urban and Rural Settlements, Carter H.	Oliver & Boyd, 1989	0 05 004286 6
The Geography of Settlements Daniel P. & Hopkinson M.	Oliver & Boyd, 1989	0 05 004286 6
An Introduction to Agricultural Geography, Grigg D	Routledge (2nd Edition), 1995	0 415 08443 1

Agricultural Change in Britain, Ilberry B.W.	Oxford Univ. Press, 1992	0 19 913365 4
The Geography of Rural Change, Ilberry B.W.	Longman, 1997	0 582 27724

Tourism and leisure

Leisure Recreation and Tourism, Prosser R.	Collins, 1994	0 00 326645 7
Critical Issues in Tourism, Shaw G. & Williams A.M.	Blackwell, 1994	0 631 18131 8

2. Magazine articles

Magazine	**Title of article and date (or number)**

General fieldwork techniques

Geographical Magazine	'Vital statistics – designing a fieldwork Project', September 1992
	'Sampling techniques for fieldwork', October 1992
	'Fieldwork techniques for infiltration studies', December 1992
	'Graphing techniques for fieldwork', December 1992
	'Don't throw away your data – making the best of your data', January 1993
	'Mapping census data', July 1993
	'Planning a personal investigation', April 2000 [in 'Teaching Geography']
Geofile	'Using and mapping census statistics', Nos 248 & 249 (Sept. 1994)
	'Conducting a fieldwork enquiry – how to record', No. 280 (Jan. 1996)

'Measuring and analysing the physical
environment', No. 283 (Jan. 1996)

'Mapping techniques and statistical
analysis for A level', No. 369 (Jan. 2000)

'Fieldwork: measuring the physical
environment', No. 380 (April 2000)

Geography Review

'Field sketching', January 1995

'Approaches to geographical enquiry',
March 1995

'Photography and field sketching in
coursework', January 1999

'Risky work', March 1999

'Quadrats: not only for biologists',
January 2000

'Dealing with variables', March 2000

'Measurement in physical geography:
the error of our ways', May 2000

'Project pitfalls', May 2000

'Fieldwork for the new millennium',
September 2000

'Realistic fieldwork', November 2000

Biogeography, ecosystems, soils, vegetation, weather and environment

Geography

'Eutrophication of lakes and
reservoirs', January 1995

'Environmental degradation and
preservation. [Humber Wetlands]',
April 2000

Geographical Magazine

'Assessing the future of National
Parks', August 1991

'Fieldwork in bogland areas'
[Hydrosere Development], September
1992

Geofile	'Changing perspectives in waste management', No. 241 (April 1994)
	'What to do with waste', No. 241 (April 1994)
	'Anticyclonic weather in the UK; its many variations', No. 355 (April 1999)
	'Wetland ecosystems', No. 362 (Sept. 1999)
Geography Review	'Monitoring water pollution', March 1991
	'Air pollution and vegetation', November 1991
	'Energy in ecosystems', September 1992
	'Soil profiles', January 1993
	'The nitrate issue', May 1993
	'Heather moorland ecosystems', November 1993
	'Investigating air quality', January 1994
	'Eutrophication', March 1994
	'Waste management – re-use and recycling', May 1995
	'Sulphur emissions and acid rain', September 1995
	'Air pollution and the nitrogen cycle', November 1995
	'Inorganic waste recycling', March 1996
	'The nature of precipitation', July 1996
	'Precipitation and the weather', October 1996

A
B
C
D
E
F
G
H
I
J
K
L
M
N
O
P
Q
R
S
T
U
V
W
X
Y
Z

'Variations in soil', March 1997

'Environmental impact assessment', November 1998

'Soil forming factors in the Isle of Man', September 1999

'Management options for Epping Forest', January 2000

'Investigating soils', March 2001

'Environmental lapse rates in the Isle of Man', November 2001

Coastal studies

Geographical Magazine

'"Sands of time" (a sand dune study)', March 1992

'Longshore drift study' [at Porlock Bay], April 1992

'The evolution of Northumberland's Aln Estuary' [mud flats and vegetation succession study], December 1993

'What caused Black Gang Landslide' [Isle of Wight coast], April 1994

'Investigating sand dunes', November 1994

'Out on a limb – erosion of Spurn Head Spit, Yorkshire', June 1996

Geofile

'Managing coastal erosion', No. 201 Sept. 1992

'Human development of coastlines', No. 273 Sept. 1995

'Basic coastal processes', No. 297 Jan. 1997

'Coastal management', No. 339 Sept. 1998

Geography Review

A
B
C
D
E
F
G
H
I
J
K
L
M
N
O
P
Q
R
S
T
U
V
W
X
Y
Z

Glaciation

Geofile	'Glaciers – form and process', No. 198 Sept. 1990
	'Periglacial landforms', No. 264 April 1995
Geography Review	'Water and ice – some key questions', September 1991
	'Glaciation and deglaciation', January 1996
	'Sculpted by ice: landforms of upland Britain', November 1999
	'Earth's giant bulldozers', May 2001

Hydrology and slopes

Geography	'The dynamics of dynamic river channels', April 1995
	'Accelerated soil erosion in Britain', April 1997
Geofile	'Human impact on drainage basin hydrology', No. 87, April 1987
	'River channels and management issues', No. 281, April 1989
	'Slope processes', No. 211, Jan. 1993
	'River channel processes', No. 251, Jan. 1995
	'New ideas in physical geography', No. 262, April 1995
	'Managing the flood hazard', No. 268, Sept. 1995
	'Catchment area – drainage basin studies', No. 316, Sept. 1997
	'Crisis studies in river management', No. 330, April 1998

Geography Review

A
B
C
D
E
F
G
H
I
J
K
L
M
N
O
P
Q
R
S
T
U
V
W
X
Y
Z

'River management and restoration',
January 2001

Rural geography: settlement, agriculture and recreation

Geography	'Farm Based Recreation', January 1996
Geographical Magazine	'National Parks – Scotland', March 1995
	'Old ways for the New Forest – management issues', March 1995
	'Leisure and the British countryside', April 1995
	'Saving the Lake District', May 1995
Geofile	'New recreational developments', No. 278, Jan. 1996
	'Changes in Rural Britain', No. 288, April 1996
	'Studies of managed environments – country parks', No. 288, April 1996
	'Human use of limestone areas', No. 302, Jan 1997
	'EU farming – issues and policies', No. 332, Jan. 1998
	'Change in UK villages', No. 325, Jan. 1998)
Geography Review	'Comparing places – villages and rural areas', May 1992
	'Fieldwork into the rural past', March 1993
	'Pleasure and environmental change', March 1995
	'England's green and pleasant land', November 1998
	'The geography of UK agriculture', November 2000

'Crisis in the countryside',
September 2001

'Second homes in England and Wales',
May 2002

Urban geography, population, industry and employment

Geography

'Development cycles and the urban
landscape', January 1994

'Whatever happened to regional
shopping centres', October 1994

'The constraints of urban history on
the form of urban redevelopment',
April 1995

'Factory outlets – shopping
developments', July 1995

'Teaching the geography of crime', July
1995 (in 'Teaching Geography')

'Re-imaging a city' [Birmingham],
January 1996

'The Development of park and ride in
Britain', January 1996

'Urban traffic congestion and
integrated transport', January 2000

'Selling the past: heritage tourism and
place identity in Stratford-upon-Avon',
July 2000

'Oxford's changing geography', July 2000

Geographical Magazine

'The geographical impact of the car',
February 1994

'Reducing the demand for car travel',
April 1994

'The superstore is back in town', June
1994

'Congestion, pollution and the cycle culture', June 1994

'Plotting the landscape – allotments in the urban field', November 1994

'The urban park – urban fieldwork techniques', December 1994

'New life in old towns – the revival of Bradford', August 1995

'Ticket to ride – a solution to urban congestion', November 1995

'Whose streets are they anyway? – reclaiming the streets from traffic', June 1996

Geofile

'British inner cities', No. 190, Jan. 1992

'Studies in the central business district', No. 220, April 1993

'Retail developments in the UK', No. 215, April 1993

'The GB census (1991)', No. 248, Sept. 1994

'Using census data', No. 249 (Sept. 1994)

'Urban study – case study of Glasgow', No 274, Jan. 1996

'Crime in Oxford', No. 287, April 1996

'Study of a UK city – Swansea', No. 293, Jan. 1997

'Services and retail change in the UK', No. 312, Sept. 1997

'Locating a new leisure and retail complex – a decision making exercise', No. 319, Sept. 1997

'Urban regeneration in London's Docklands', No. 329, April 1998

'Inner city policies – 1945–1998', No. 336, Sept. 1998

'Oxford's green belt', No. 337, Sept. 1998

'Changing location of offices - Birmingham case study', No. 340, Sept. 1998

'Inner city redevelopment in Manchester', No. 351, April 1999

'Suburbanisation: London case study', No. 353, April 1999

'London Docklands: an update', No. 357, April 1999

Geography Review

'Carrying out an industrial study', January 1991

'Investigating industrial estates', May 1991

'Exploring the past' [Case study of Bath], November 1992

'The changing geography of retailing', January 1994

'Geography of ethnic minorities', March 1994

'The geography of economic activity', May 1994

'Energy, climate and retail change – a study of the impact of supermarkets', May 1994

'The changing manufacturing base of the United Kingdom', September 1994

'The geography of crime and crime prevention', March 1996

'Suburbanisation', September 1998

A B C D E F G H I J K L M N O P Q R S T U V W X Y Z

'Urban tourism in the 1990s: understanding the figures for London', November 1998

'Regeneration in England's inner cities', January 1999

'Make your own urban case study', March 1999

'Deprivation and health in Sheffield', November 1999

'Footloose services: the economic geography of telephone call centres', September 2000

'Urban regeneration', May 2002

Urban physical

Geography Review

'Investigating urban weather' [especially gravestones], March 1992

'The impact of acid rain on buildings', January 1993

'Urban fieldwork and physical geography', May 1997

Geofile

'Urban biogeography', No. 326, Jan. 1998

A-Z of useful addresses

1. Census and other data

Name and address

Census Customer Services (England/Wales)
Segensworth Road
Titchfield
Hampshire PO15 5RR

Census Customer Services (Scotland)
General Register House
Ladywell House
Edinburgh EH12 7TF

HMSO
PO Box 276
London SW8 5DT

Department for the Environment, Food and Rural Affairs (DEFRA)
(Publications)
Admail 6000
London SW1A 2XX

Office of Population, Censuses and Surveys
St Catherine's House
10 Kingsway
London WC2 6JP

Office for National Statistics – Library
1 Drummond Gate
London SW1V 2QQ

Office for National Statistics – Sales
The Stationery Office
PO Box 276
London SW8 5DT

Public Record Office
Ruskin Avenue
Kew
Richmond
Surrey TW9 4DU

The Countryside Commission
John Dower House
Crescent Place
Cheltenham GL50 3RA

2. Conservation

Countryside Council for Wales
43 The Parade
Roath
Cardiff CS2 3UH

English Nature
Northminster House
Peterborough PE1 1UA

Field Studies Council
Preston Mountford
Mountford Bridge
Shrewsbury
Shropshire SY4 1HW

Friends of the Earth
26-8 Underwood Street
London N1 7JQ

Greenpeace
Temple House
25-6 High Street
Lewes
East Sussex BN2 7LU

National Society for Clean Air and Environmental Protection
136 North Street
Brighton BN1 1RG

Royal Society for Nature Conservation
The Green
Witham Park
Waterside South
Lincoln LN5 7JR

Scottish National Heritage
12 Hope Terrace
Edinburgh EH9 2AS

Acid Rain Information Centre
Manchester Metropolitan University
All Saints
Manchester M15 6BH

3. The environment

British Geological Survey
Kinsley Dunham Centre
Keyworth
Nottingham NG12 5GG

Council for Environmental Education
University of Reading
London Road
Reading RG1 5AQ

The Forestry Commission
231 Corstophine Road
Edinburgh EH12 7AT

Geographical Association
343 Fulwood Road
Sheffield S10 3BP

Geologists' Association
Burlington House
Piccadilly
London W1V 0JU

Meteorological Office
Education Service
Johnson House
London Road
Bracknell
Berkshire RG21 2SZ

National Rivers Authority
Rivers House
Waterside Drive
Aztec West
Bristol BS12 4UD

National Water Archive
Institute of Hydrology
Maclean Building
Crowmarsh Gifford
Wallingford
Oxfordshire OX10 8BB

WATCH
The Green
Witham Park
Waterside South
Lincoln LN5 7JR

4. Maps and photographs

Aerofilms (Aerial photographs)
Gate Studios
Station Road
Borehamwood
Herts WD6 1BR

Geological Museum (Geological maps)
Exhibition Road
London SW7 2DE

Goad Maps (Goad maps)
8-12 Salisbury Square
Old Hatfield
Herts AL9 5BJ

Land-use Research Unit (Land-use maps)
King's College
University of London
The Strand
London WC2R 2LS

National Map Centre (O.S. Maps)
22 Caxton Street
London SW1H 0QH

National Remote Sensing Centre (LANDSAT and satellite images)
Customer Services
Delta House
Southwood
Farnborough
Hampshire GU14 8JY

Ordnance Survey (O.S. Maps)
Romsey Road
Southampton SO9 4DH

Soil Survey and Land Research Centre (Soil maps)
Publications Officer
Silsoe Campus
Silsoe
Bedfordshire MK45 4DT

5. Others

Association of British Chambers of Commerce
9 Tufton Street
London SW1P 3QB

Confederation of British Industry
Centrepoint
103 New Oxford Street
London WC1A 1DU

Market Research Society
15 Northburgh Street
London EC1V 0AH

Trades Union Congress
Congress House
Great Russell Street
London WC1B 3LS

A-Z of geography websites

Websites by their very nature can change, move or become outdated. There are a number of sites here that you could consult, but remember that some of these suggestions are merely a possible starting point for further research.

Web address	Content
http://galaxy.einet.net/galaxy/social-sciences/geography/	The Einet search engine lists geographical information under the titles topics, books, software, cartography, collections, directories and organisations.
http://www.zephyrus.demon.co.uk/geography/home.html	The Geography Exchange site has links to sites, sorted by topic.
gopher://envirolink.org:70/1/	The Envirolink organisation's home pages provide links to a range of information, issues, publications and organisations concerned with environmental protection.
http://www.utexas.edu/depts/grg/virtdept/contents.html	Virtual Geography department
http://Yahoo.com/regional/countries/	A starting point for websites about different countries.
http://the-times.co.uk	An electronic edition of The Times newspaper. Contains an archive search for past articles.
http://www.stockportmbc.gov.uk/pages/links/curricul/geog/geo4.htm	Provides many links
http://www.newscientist.com	Magazine
http://vtc.ngfl.gov.uk	Virtual teacher centre
http://www.volcano.und.nodak.edu/	Sponsored by NASA, this site has myriad data, interactive maps and hundreds of pictures of volcanoes throughout the world.

http://www.cla.org.uk	Country Landowners' Association
http://www.english-nature.org.uk	Environment
http://www.open.gov.uk/	Environment
http://www.tourism.wales.gov.uk	The Welsh Tourist Board world facts and figures
http://educeth.ethz.ch/stromboli/index-e.html	Virtual fieldtrip up Stromboli
http://www.odci.gov/cia/publications/94fact/fb94toc/fb94toc.html	Tourism
http://sunsite.unc.edu:80/world/wordhome.html	Tourism
http://coombs.anu.edu.au/resfacilities/demographypage.html	Demography and population
http://sosig.ac.uk/subjects/demog.html	Demography and population
http://www.kuntalitto.fi/hallinto/toetoh/gis.html	GIS (Geographical Information Systems)
http://www.geo.ed.ac.uk/hom/gishome.html	GIS (Geographical Information Systems)
http://www.metro2000.net/~stabbott/sepnat.htm	Separatism
http://www.pci.on.ca/rssites.html	Remote Sensing
http://www.vtt.fi/aut/aua/rs/virtual	Remote Sensing
http://osiris.wu-wien.ac.at/regsci/.regsci.html	Regional Science
http://www.echo.lu/other/otherhome.html	Government – this site has links to 18 different countries' government servers
http://www.rdg.eurobell.co.uk/QUODITCH-html	Biogeography of a hydrosere
http://www.onelineweather.com	Weather
http://www.bbc.co.uk/weather	Weather
http://www.met-offfice.gov.uk	Weather

A
B
C
D
E
F
G
H
I
J
K
L
M
N
O
P
Q
R
S
T
U
V
W
X
Y
Z

http://www.atm.ch.cam.ac.uk/	Weather
http://rs560.cl.msu.edu:80/weather/	Weather
http://www.meto.govt.uk	The UK Met office. Provides daily weather satellite images and has a student's enquiry page.
http://lal.k12.ca.us/aep/start/river/tour/index/html	This site looks at flood defences, pollution and other issues concerning a Los Angeles river.
http://www.le.ac.uk/cti/	The CTICG (Computers in Teaching Initiative, Centre for Geography) contains information concerning its own geographic computer-assisted learning software, plus it provides links to information concerning human and physical geography, cartography, environment resources, GIS and remote sensing, geology and place information.
http://www.yahoo.com/science/geography	A general geography site
http://ecosys.drdr.virginia.edu/environment.html	Environment
http://www.yahoo.com/environment_and_nature/	Environment and Nature
http://www.ran.org/ran/	This site provides case studies of rainforest issues, especially ecotourism, and links to other websites.
http://www.environment-agency.gov.uk/start.page.shtml	Environment Agency
http://www.aeat.com./netcen/airqual/welcome.hmtl	Local air quality site that enables you to see how your local authority is performing. Also indicates if an authority has designated air quality management zones where driving can be restricted if air quality is poor.
http://www.aecw.demon.co.uk/rrc/rrc.htm	River Restoration Centre gives information on river projects.
http://www.pap-thecoastcentre.org	Information on coastal acton with preference to the Mediterranean

http://www.geocities.com/geographyatbgs/new/Maidstone.htm	Virtual urban transect
http://www.library.thinkquest.org/C003603/english/flooding/causesoffloods.shtml	Types of floods
http://www.yorkshirenet.co.uk/visinfo/ydales/malham.htm	Malham Cove and limestone scenery
http://www.brighton.ac.uk/environment/ROCC/sussex.htm	Cliff erosion at Beachy Head
http://www.pml.ac.uk/lois/Education/eros.htm	Coastal study at Holderness
http://www.agri.upm.edu.my/jst/soilinfo/html	General soil site
http://www.defra.gov.uk/	Department for the Environment, Food and Rural Affairs (DEFRA), formerly MAFF
http://www.countryside.gov.uk/	Countryside Agency
http://www.ruralnet.org.uk/	The future of rural England

A
B
C
D
E
F
G
H
I
J
K
L
M
N
O
P
Q
R
S
T
U
V
W
X
Y
Z